RAND NATIONAL DEFENSE RESEARCH INSTITUTE

A Review of Alternative Methods to Inventory Contracted Services in the Department of Defense

Nancy Y. Moore, Molly Dunigan, Frank Camm, Samantha Cherney, Clifford A. Grammich, Judith D. Mele, Evan D. Peet, Anita Szafran

Prepared for the Office of the Secretary of Defense

For more information on this publication, visit www.rand.org/t/RR1704

Library of Congress Cataloging-in-Publication Data
ISBN: 978-0-8330-9672-2

Published by the RAND Corporation, Santa Monica, Calif.

Cover: Fotolia LLC/Syda Productions.

Preface

Title 10, Section 2330a, of the U.S. Code requires the Secretary of Defense to "submit to Congress an annual inventory of the activities performed during the preceding fiscal year pursuant to contracts for services." Persistent concerns regarding both the methods for collecting these data in the Inventory of Contracted Services (ICS) and the utility of the data led the conferees for the National Defense Authorization Act for Fiscal Year 2016 to direct the Secretary of Defense to examine the approach that the U.S. Department of Defense (DoD) is taking to comply with this statutory requirement. Congress directed the Secretary of Defense, as part of this examination, to determine whether the ICS produced by DoD enhances oversight of contracting activities and to submit a report to the congressional defense committees explaining the results of that examination, outlining efforts to better manage contractor and civilian personnel costs within DoD, and outlining potential alternative methods of meeting ICS requirements.

To assist the Secretary of Defense in making this determination, the Principal Deputy Assistant Secretary of Defense for Manpower and Reserve Affairs asked the RAND Corporation to conduct the mandated research. This final report builds on an interim report delivered in advance of the March 1, 2016, deadline for reporting to Congress. It should be of interest to policymakers concerned with DoD purchases of services, as well as to DoD officials charged with ensuring better oversight of purchased services.

This research was sponsored by the Principal Deputy Assistant Secretary of Defense for Manpower and Reserve Affairs and conducted within the Forces and Resources Policy Center of the RAND National Defense Research Institute, a federally funded research and development center sponsored by the Office of the Secretary of Defense, the Joint Staff, the Unified Combatant Commands, the Navy, the Marine Corps, the defense agencies, and the defense Intelligence Community. For more information on the RAND Forces and Resources Policy Center, see www.rand.org/nsrd/ndri/centers/frp.html or contact the director (contact information is provided on the web page).

Contents

Figures and Tables

Figures

Tables

Summary

Since the late 1940s, U.S. Department of Defense (DoD) purchases of services have increased consistently, from less than 30 percent to more than 60 percent of the department's overall budget. This increase reflects both the growth of services in the overall economy and the initiatives of political administrations over time to procure services from the private sector on behalf of DoD to the greatest extent possible. Nevertheless, such growth has led to concerns regarding contracting of inherently governmental functions, contract oversight, contractor accountability, and contract waste, fraud, and abuse.

Concerns about the growth in DoD's purchases of services have led Congress to institute several policies aimed at strengthening oversight of such purchases. These policies have included 2001 legislation requiring DoD to collect and track data on the procurement of services, 2002 and 2008 congressional language expressing an interest in spend analyses that might be used to increase buying leverage and improve contractor performance, and a 2008 requirement in Title 10, Section 2330a, of the U.S. Code establishing the DoD Inventory of Contracted Services (ICS) to collect information on activities performed under DoD service contracts.

Concern regarding both the methods for collecting data in the ICS and the utility of these data led Congress to request that the Secretary of Defense review the methods used to create the ICS, as well as the products resulting from these efforts. Congress specifically requested that the Secretary of Defense examine the extent to which the ICS provides data on service contracts that are useful to DoD and congressional stakeholders, the extent of gaps between ICS data and data that DoD and Congress would find most useful, whether existing databases or other information technology systems could provide a timely solution and data that are relevant to workforce planning, and the strengths and weaknesses of different methods for reporting on DoD's use of contractor personnel. DoD asked RAND to assist the Secretary of Defense in fulfilling this congressional mandate.

This report documents the final results of that research. It explores the congressional intent underlying the ICS requirement, gaps between the ICS data and data most useful to DoD and congressional stakeholders, insights on the issues that Congress seeks to address through the ICS requirement that can be derived from analyses of non-ICS data found in alternative databases, and the strengths and weaknesses of different methods for estimating and reporting contractor personnel use.

Research Methods

This study employed multiple research methods and was conducted in a compressed time frame. The bulk of the data collection and analysis was completed between mid-December 2015 and mid-February 2016 to produce an interim report in advance of the Secretary of Defense's March 1, 2016, deadline for reporting to Congress. During that time, we reviewed relevant legislation and literature; analyzed relevant data from the ICS, the Federal Procurement Data System—Next Generation (FPDS-NG), and the System for Award Management (SAM); and interviewed key stakeholders in Congress, DoD, non-DoD federal agencies, and the offices of relevant service contractors. Over the course of the project, we interviewed 83 individuals and reviewed more than 80 documents, focusing on the legislative and historical context underlying the ICS, as well as insights from the economics literature. We also analyzed ICS and FPDS-NG data to develop distribution and trend data on spending, contracts, business size, and type of service, as well as to identify contractors to interview. Finally, we devised and tested several alternative metrics for calculating contractor full-time equivalents (FTEs) using existing non-ICS data sources.

What Does the Current ICS Look Like?

The current ICS is produced approximately one year after the end of the fiscal year (FY) for which data are reported and is captured in two publicly available formats: a report to Congress and 37 different defense-component spreadsheets on the Defense Procurement Acquisition Policy (DPAP) website. The ICS is produced using the Contractor Manpower Reporting Application (CMRA) system. The Army first developed the CMRA system, but now there are four separate "instances," or versions, of the system—one each for the Army, Air Force, and Navy and a combined one for the other defense agencies. As currently planned, the different instances of CMRA will be combined into one "enterprise-wide" system (eCMRA) in the next several years, and all instances are now being moved under Defense Manpower Data Center stewardship.

We were unable to gain access to the raw CMRA data for this study, as access is limited in an attempt to protect contractors' proprietary data from competitors. However, it is critical to note that even without access to restricted CMRA data, we were able to link ICS-reported direct labor hours to particular service contractors using contract number information publicly available on the FPDS-NG website and the publicly available ICS data published on the DPAP website (which reports contract number as well as direct labor hours information). When we analyzed the ICS data and compared them to FPDS-NG data, we also found shortfalls in completeness and quality, which are discussed in Appendix C of this report. These analyses reinforce some of what we heard in our interviews with various stakeholders and subject-matter experts.

How Well Does the ICS Meet Congressional Objectives and DoD Needs?

In our interviews with congressional staff and DoD stakeholders, we found that the current ICS falls short of meeting the needs of Congress and DoD. Many congressional staff suggest that the format in which ICS data are reported to Congress is not useful and hinders assessment of the data. Several commented that the data, as reported, are too detailed and would be more useful if they were synthesized before reporting. Ultimately, it appears that Congress seeks analysis—not raw data—from DoD, but this is not well specified in the statute.

The views of DoD stakeholders, meanwhile, vary based on the interests of their functional communities. Manpower and personnel, budgeting, and acquisition officials require different information to do their jobs most effectively. This, in turn, shapes their views of the utility of the ICS. Stakeholders who focus on manpower and personnel planning, for example, seek data on contractor FTEs and level of effort needed to enable strategic workforce planning and insourcing decisions. Those in the budgeting community seek data on total costs and data that integrate well into budget considerations, allowing them to budget more effectively. Meanwhile, those in the acquisition community seek data on level of performance and total costs to enable smart acquisition decisionmaking. Such variation in the preferred types of data on service contracts makes it difficult to determine what data need to be collected and why. Understanding the goals of collection is critical in making this determination.

The characteristics and types of data that appear to be most relevant to congressional and DoD stakeholders are (1) processed, analyzed data; (2) forward-looking data that can be integrated into budget processes; (3) data on contractor FTEs to compare with civilian FTEs in making sourcing decisions; (4) auditable and verifiable data; and (5) data distinguishing types of contracts by total costs, contractor FTEs, and other values of interest. By contrast, the ICS includes data that are unprocessed, retrospective, and can largely be found elsewhere, with the exception of contractor direct labor hours. Moreover, the direct labor hours data included in the ICS were, at the time this research was conducted, largely estimated rather than contractor-reported, making them difficult to verify or even distinguish among contracts.

Meanwhile, our interviews with service contractors indicated that CMRA reporting can be burdensome for the contractor and that contractors are subject to a multiplicity of reporting requirements, some mandating that they enter overlapping data points into CMRA and other systems, such as SAM. Moreover, contractors questioned the utility of collecting direct labor hours data and were concerned about the exposure of their proprietary data and how that may affect their success in competing for future contracts.

Why Are There Gaps Between the Current ICS and What Congress and DoD Envisioned?

To understand the shortcomings of the ICS and the challenges in meeting congressional intent related to the ICS requirement, it is critical to note that service contractors' production functions vary, so comparing metrics across these firms can be misleading. Yet the ICS is structured to measure contractors using equivalent inputs, as though they all produce equivalent services. This has the potential to distort results, as there is extensive variability between service contractors in the *types* of services they provide and, particularly, the degree to which the services they provide replace or simply *augment* governmental functions. Furthermore, service contractors demonstrate great variability in how they produce outputs, specifically in terms of the degree to which they substitute capital for labor and their various types of labor input. Indeed, collected labor input data show that although direct labor accounts for about half of total contract costs, the direct labor fraction varies greatly by type of service, from about one-fourth to three-fourths of total costs.

Table S.1 illustrates the spectrum of contracting activities in which DoD may engage, ranging from staff augmentation contracting (also known as "labor contracting") to complete contracting, with mixed contracting lying between the two extremes. In instances of staff augmentation contracting, DoD provides the facilities, materials, equipment, technologies, and other inputs to production. Meanwhile, in complete contracting, DoD provides only contractor management. Because of the distinction in how these levels of contracting are managed, collecting direct labor hours for all DoD service contracts without distinction in terms of the types of services provided is problematic. Even assuming that data on direct labor hours are valid and precise, collecting them for complete contracting is inappropriate because each contractor engaged in complete contracting makes distinct decisions regarding the inputs, processes, and practices used to provide the service. Because direct labor hours do not account for distinctions between the various types of contracting activities, they are insufficient to inform strategic workforce planning or DoD budget decisionmaking and acquisition planning.

Exacerbating the insufficiency of direct labor hours for informing strategic workforce planning is the fact that substitutions between different components of the total force—military, civilian, and contractor—cannot always be exchanged one-for-one within and across sectors because of individual-, organization-, and sector-level variations and gaps in productivity. For instance, different organizations tend to hire workers from different backgrounds, motivate them in different ways, and train them to have different skill sets using distinct methods. Maximizing labor productivity would clearly be ideal. However, without precise measures of productivity, and with legal constraints on sourcing decisions and governmental influence in contractor labor decisions—such as a moratorium on outsourcing competitions and constraints on military and civilian personnel hiring—the ability to use proxy measures of productivity correctly and appropriately is key to informing strategic workforce management. The collection of direct labor hours in the ICS is *not* an appropriate proxy measure of productivity, especially when these data make no distinction between the various types of contracting activities being performed.

Table S.1
Distinct Contracting Activities Require Different Management

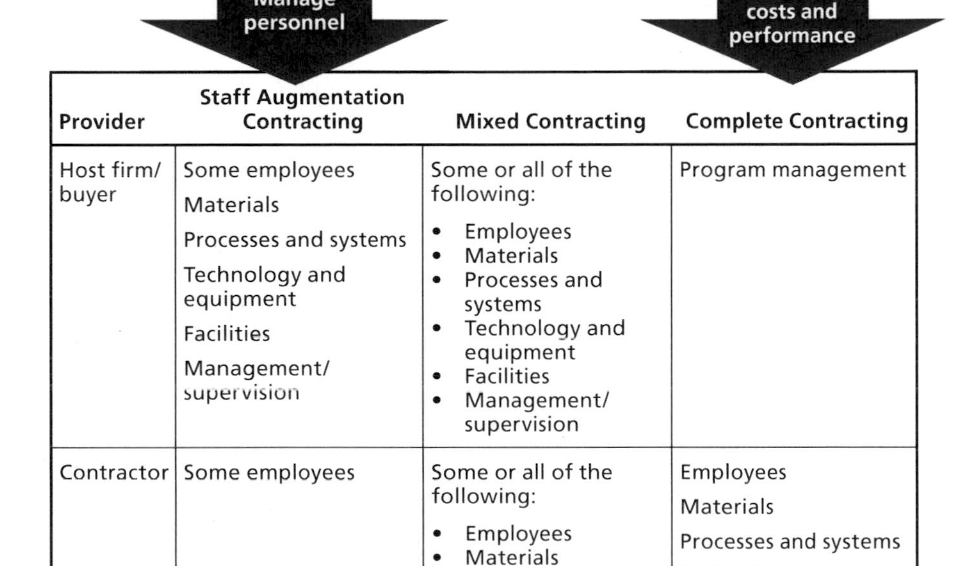

Provider	Staff Augmentation Contracting	Mixed Contracting	Complete Contracting
Host firm/ buyer	Some employees Materials Processes and systems Technology and equipment Facilities Management/ supervision	Some or all of the following: • Employees • Materials • Processes and systems • Technology and equipment • Facilities • Management/ supervision	Program management
Contractor	Some employees	Some or all of the following: • Employees • Materials • Processes and systems • Technology and equipment • Facilities • Management/ supervision	Employees Materials Processes and systems Technology and equipment Facilities Supervision

SOURCE: Adapted from Sandy Allen and Ashok Chandrashekar, "Outsourcing Services: The Contract Is Just the Beginning," *Business Horizons*, Vol. 43, No. 2, March 2000.

Insights on DoD Service Contracting Provided by Data Systems Other Than and the ICS

Our work exploring the potential to meet congressional intent for the ICS with the use of other data systems focused primarily on data from the FPDS-NG (and, to a lesser extent, on budget data). While FPDS-NG data may contain some errors in data submission, it is the authoritative system for federal contract reporting, and the quality of its data has improved over time. FPDS-NG provides, for contract actions of at least $3,000, information on the amount of the contract action, identification codes indicating whether the firm providing the service is a small business, the North American Industry Classification System (NAICS) code for the firm, the Treasury Account Symbol for the transaction funding (which can be linked to budget categories), and the Product or Service Code (PSC), a more finely grained indicator than the NAICS code regarding the exact nature of goods and services purchased. Though subject to some delay in publication due to security measures and verification, these data can provide numerous insights on the services DoD has recently purchased and, in doing so, can assist in addressing the various congressional concerns underlying the ICS requirement—namely, enabling the production of spend analyses, trend analyses, and forecasting to inform budgeting and acquisition decisions. As we discuss in greater detail in the next section, FPDS-NG data can also be used to produce alternative metrics for calculating contractor FTEs in an effort to inform strategic workforce planning.

In terms of their contribution to spend and trend analyses, FPDS-NG data indicate that half of DoD service spending falls under three PSC categories: Support (Professional/Administrative/Management), Research and Development, and Maintenance, Repair, and Rebuilding of Equipment. Further probing of FPDS-NG data shows that four specific types of services—including engineering and technical services and general health care services—were significant drivers of increases in DoD support service spending. FPDS-NG data indicate some opportunities to leverage purchases (that is, to consolidate contracts or purchases across offices so as to increase buying power), but they also point to possible difficulties in doing so. These potential challenges include the large proportions of small businesses and the wide array of industries (denoted by NAICS codes) providing these services, each of which is likely to vary along a number of dimensions. Finally, FPDS-NG data also help to illustrate the extent to which current service purchases are open to competition, as well as the contract types used to purchase services.

Coupling FPDS-NG data with budget-category projections can yield insights regarding likely future trends in overall spending for services. Most spending (59 percent) for services is related to operations and maintenance (O&M), one of the categories Congress uses for budgeting. Current budget projections indicate that O&M spending will continue to decrease, meaning spending on contracted services is likely to decrease as well. Congress stated that it wanted DoD to achieve a reduction in service spending of $4.1 billion by FY 2017, relative what it was spending in FY 2012 ($186 billion). This amount of reduction in services spending—$4.1 billion—is equivalent to a parallel reduction in military basic pay resulting from reductions in military end strength in the same period. Calculating actual spending reductions using FPDS-NG data indicated that DoD had already more than met this goal in FY 2015, reducing service spending by $38 billion. Using the President's budget projections, and assuming that DoD out-year spending matches these budget projections and a constant percentage use of service spending occurs in each budget category over time, we estimated that the reduction in service spending will continue along the same trend, decreasing by $60 billion between FY 2012 (when total service spending was $186 billion) and FY 2021 (when we project total service spending to be $126 billion).

Risks and Benefits of Different Methods for Estimating and Reporting Contractor Personnel Use

In our interviews, we found that one of the key motives underlying the collection of data on direct labor hours associated with a contract is to use this information to assess the scale of the contracted services relative to the size of comparable DoD in-house activities. However, due to the shortcomings of relying on direct labor hours data for strategic workforce planning and insourcing decisions, as discussed earlier, DoD might consider alternative measures that do not require collecting, validating, auditing, and protecting proprietary data reported by contractors.

We identified three alternative metrics to estimate contractor manpower numbers, in addition to the current ICS metrics (both actual contractor-reported direct labor hours and direct labor hours calculated using Army algorithms that are based on previously reported data on firms providing similar services). These are as follows:

1. the number of civilian FTEs that could be hired with the contract dollars ("civilian labor FTE per contract")
2. the number of industry or location-average employees per contract dollars ("contractor labor FTE per contract")
3. contract employees as a proportion of overall contractor revenue.

These metrics may be calculated from data available through FPDS-NG, the Bureau of Labor Statistics, and the U.S. General Services Administration–owned SAM, which consolidates the Catalog of Federal Domestic Assistance and various federal procurement systems.

Because these alternatives draw on available, in-house federal data or publicly available data, they do not require DoD to collect, validate, audit, and protect proprietary data from contractors as the current metrics do. This, in turn, would likely generate cost savings, as the expenses incurred by contractors to collect and report direct labor hours on a given contract are included in the overall price of that contract. The use of these alternative metrics in lieu of contractor-reported or estimated direct labor hours could also assist DoD in producing an ICS in a more timely manner, as they might not be as time-consuming to generate. The common disadvantage of these three alternative metrics is that they assume equal productivity across employees, industries, and sectors. Nevertheless, our comparative analyses of the results of the current ICS-derived metrics and these alternative metrics for determining the relative importance of contracted versus noncontracted labor across functions—based on calculations performed using each respective metric on "case studies" of particular PSCs—indicate that these alternatives are close proxies for the ICS metrics.

Conclusions

Our findings suggest that the ICS products, and the processes used to create them, are not meeting either congressional or DoD stakeholder needs. Several factors led us to this conclusion. First, the congressional intent underlying the ICS requirement is multifaceted and not always clearly specified in statute. Second, different ICS stakeholders are based in distinct functional communities, each of which has its own interests and needs driving its purpose for utilizing ICS data—and these needs and purposes do not always align across these divergent communities. Third, opinions differ both inside and outside of DoD on the utility and quality of the current ICS data, with some stakeholders finding the data more valuable and some finding them less valuable. Fourth, because the majority of ICS data through FY 2014 (the most recent year for which ICS data were available during the period of research) are derived using algorithms developed by the Army that are based on unverified contractor-reported data, their validity is questionable from the outset—particularly for contracts held by military services and defense components other than the Army. Moreover, the ICS data do not currently support spend analyses, trend analyses, forecasting, or strategic sourcing, and more information would be needed to conduct effective labor comparisons to inform insourcing decisions. Finally, much of the information Congress seeks to allow oversight of service contracts is available in other systems.

These findings led to several recommendations. First, policymakers should institutionalize the development and reporting of DoD-wide spend analyses of services, including analyses of trends, forecasts, and FTEs. This would entail issuing a detailed requirement for an institutionalized capability to analyze data on DoD service contracts and providing the necessary funding for its development. DoD would also likely need to employ dedicated research programmers or statistical analysts in long-term positions to produce ICS-related analyses.

Second, ICS-related statutory requirements could be refined to better distinguish between different types of contracting and, accordingly, to require the collection of different data elements for each. Our research found that DoD contracting practices vary with both the types of services purchased and the level of oversight DoD expects over such purchases. ICS requirements could be revised to identify and distinguish among staff augmentation, mixed contracting, and complete contracting arrangements. For staff augmentation contracts, ICS requirements could be revised to specify the use of multiple alternative metrics relying on existing data sources, such as the FPDS-NG, to estimate a likely range of contractor FTEs. For mixed and complete contracting, the ICS requirement could be rewritten to focus on measuring total cost and performance, rather than direct labor hours. Finally, for operational support contracts—for which Congress wants increased oversight of the number of deployed contractors on the ground—reporting requirements should focus on the number of actual deployed contractors, not FTEs.

Third, DoD should periodically perform sourcing analyses of selected commercial services to determine whether civilians or contractors deliver the required level of performance at the lowest total costs. Doing so will ensure continuous adjustment of task assignments across the total force, where necessary, to maintain the lowest cost and most effective staffing solutions for a diverse set of defense functions.

Acknowledgments

We gratefully acknowledge the assistance of a number of congressional staff members and individuals across the U.S. Department of Defense, U.S. military services, other U.S. government agencies, service contracting firms, and policy research organizations who took the time to speak with us. We also extend our gratitude to Colonel Andrea Miller and to Jayme Fuglesten and Kurt Card in RAND's Office of Congressional Relations for their assistance arranging interviews with congressional staff. At RAND, we thank John Winkler, Lisa Harrington, and Laura Baldwin for their management support throughout the course of the study, and Ed Keating, Sheila Earle, Susan Gates, Sarah Meadows, and Craig Bond for their careful reviews of this report. The Office of the Assistant Secretary of Defense for Manpower and Reserve Affairs deserves special thanks for funding this research, and within this office, we would especially like to thank Principal Deputy Assistant Secretary of Defense Stephanie Barna and Dave Sheridan.

Abbreviations

CMRA	Contractor Manpower Reporting Application
DoD	U.S. Department of Defense
DPAP	Defense Procurement Acquisition Policy
eCMRA	Enterprise-Wide Contractor Manpower Reporting Application
FPDS-NG	Federal Procurement Data System—Next Generation
FTE	full-time equivalent
FY	fiscal year
GAO	U.S. Government Accountability Office (after July 7, 2004); U.S. General Accounting Office (before July 7, 2004)
GDP	gross domestic product
GFEBS	General Fund Enterprise Business System
ICS	Inventory of Contracted Services
IT	information technology
NAICS	North American Industry Classification System
NDAA	National Defense Authorization Act

O&M	operations and maintenance
OMB	Office of Management and Budget
OSD	Office of the Secretary of Defense
PSC	Product or Service code
QCEW	Quarterly Census of Employment and Wages
RDT&E	research, development, testing, and evaluation
SAM	System for Award Management
TDA	Tables of Distribution and Allowances
UIC	unit identification code
U.S.C.	U.S. Code

CHAPTER ONE

Introduction

From 1947 to 2015, the service sector grew from 47 percent to 68 percent of U.S. gross domestic product (GDP).[1] Mirroring this trend in the overall economy, U.S. Department of Defense (DoD) purchases—including for services—also grew relative to DoD spending on military and civilian personnel during this period. For instance, in fiscal year (FY) 1948, spending on purchases accounted for less than 30 percent of DoD spending, with personnel spending totaling more than 70 percent of the department's budget. By FY 2015, spending on purchases represented 64 percent of DoD's budget, with personnel spending making up 36 percent.[2] (See Appendix A for more detail on changes in U.S. and DoD service spending over time.)

The growth in DoD's service-related spending rests, in part, on broad economic trends but also on the evolution of federal service contracting policy since the 1950s. In 1955, the Bureau of the Budget (the predecessor to the Office of Management and Budget, or OMB) stated that "the Federal Government will not start or carry on any commercial activity . . . for its own use if such product or service can be procured from private enterprise."[3] This policy was increasingly broadened

---

[1] Bureau of Economic Analysis, "Gross-Domestic-Product-(GDP)-by-Industry Data," web page, last updated October 16, 2016a.

[2] Office of the Under Secretary of Defense, Comptroller, *National Defense Budget Estimates for FY 2017*, Washington, D.C., March 2016, Tables 6-1 and 6-2.

[3] Cited in Office of the Under Secretary of Defense for Acquisition and Technology, *Report of the Defense Science Board Task Force on Outsourcing and Privatization*, Washington, D.C., August 1996, p. 3A.

in practice over time, with the 1980s "Reagan Revolution" advocating government outsourcing to increase efficiency, the 1996 Defense Science Board Task Force on Outsourcing and Privatization espousing the outsourcing of support services (except those that are inherently governmental, that directly affect warfighting capability, or for which private-sector capability is not available), and then–Secretary of Defense Donald Rumsfeld's 2001 initiative emphasizing the use of the private sector.[4] More broadly in the federal government, OMB Circular A-76, first issued in 1966, requires the periodic review of commercial functions performed internally with subsequent contracting as appropriate. While the late 1980s and early 1990s saw an effective moratorium on awarding contracts resulting from A-76 competitions, DoD gave renewed emphasis to them after this period.[5]

These initiatives culminated in the early 2000s, when growth in service contracts increased rapidly due to the unprecedented extent to which operational support functions were contracted out during Operation Iraqi Freedom.[6] Although operational contract support services account for only a portion of the total range of services for which DoD contracts, their high visibility during the conflicts in Iraq and Afghanistan, as well as numerous reports detailing contractor-related incidents in theater and multiple instances of documented waste, fraud, and abuse associated with such contracts during these conflicts, gained

---

[4] CNN, "The Reagan Years: Reaganomics," 2001; William Greider, "The Education of David Stockman," *Atlantic Monthly*, December 1981.

[5] For more on DoD and A-76 competitions, see, for example, Susan M. Gates and Albert A. Robbert, *Personnel Savings in Competitively Sourced DoD Activities: Are They Real? Will They Last?* Santa Monica, Calif.: RAND Corporation, MR-1117-OSD, 2000; U.S. General Accounting Office, *Defense Management: DOD Faces Challenges Implementing Its Core Competency Approach and A-76 Competitions*, Washington, D.C., GAO-03-818, July 2003; and Valerie Bailey Grasso, *Defense Outsourcing: The OMB Circular A-76 Policy*, Washington, D.C.: Congressional Research Service, June 30, 2005.

[6] Indeed, contractors hired by DoD at times outnumbered U.S. troops on the ground in Iraq, with 155,826 contractors working alongside 152,275 U.S. troops in theater as of 2008 (University of Denver, Private Security Monitor, "Articles, Reports, and Statistics," web page, undated).

congressional attention.[7] This high visibility—along with concerns regarding the overall growth in DoD's spending on service contracts, increasing reliance on contractors to perform staff augmentation roles, and the discovery that some contractors were performing tasks that were inherently governmental—led to congressional efforts to gain greater visibility into DoD's service contracting activities.[8]

Title 10, Section 2330a(c), of the U.S. Code (U.S.C.) requires the Secretary of Defense to

> submit to Congress an annual inventory of the activities performed during the preceding fiscal year pursuant to contracts for services (and pursuant to contracts for goods to the extent services are a significant component of performance as identified in a separate line item of a contract) for or on behalf of the Department of Defense.[9]

The statutory requirement that led DoD to establish its annual Inventory of Contracted Services (ICS) is one critical aspect of congressional efforts to gain visibility and oversight into DoD's service contracting activities. Interviews conducted for this study have elucidated that, in drafting this requirement, Congress had multiple goals: (1) to enable DoD to generate spend analyses that would allow better and more finely tuned control over the costs of service contracting; (2) to track the extensive amount of staff augmentation contracting

---

[7] See, for instance, James Glanz and Alissa J. Rubin, "From Errand to Fatal Shot to Hail of Fire to 17 Deaths," *New York Times*, October 3, 2007a; Commission on Wartime Contracting in Iraq and Afghanistan, *Transforming Wartime Contracting: Controlling Costs, Reducing Risks*, final report to Congress, Washington, D.C, August 2011; U.S. Senate, Committee on Appropriations, "Examining the Effectiveness of U.S. Efforts to Combat Waste, Fraud, Abuse, and Corruption in Iraq," Senate Hearing 110-673, March 11 and July 23, 2008.

[8] As discussed further in Chapter Four, DoD engages in various types of contracting, ranging from staff augmentation contracting to complete contracting. In staff augmentation contracting (also known as labor contracting), DoD provides the facility, materials, equipment, system, technology, and other inputs to the productive process. By contrast, in complete contracting, the contractor provides all productive inputs while DoD provides only contract management. Between these two extremes, there are many opportunities for "mixed contracting" that are negotiated and modeled to fit the circumstances.

[9] 10 U.S.C. 2330a(c).

that had been occurring by comparing the number of contractor and civilian full-time equivalents (FTEs) that could theoretically be calculated using contractors' direct labor hours; and, (3) as mentioned, to gain visibility into the numbers of deployed contractors working on DoD contracts in theaters of conflict.

However, these three distinct goals were not clearly articulated in 10 U.S.C. 2330a(c). Perhaps as a result, much of the focus of DoD's efforts has been on collecting direct labor hours for *all* service contracts above the simplified acquisition threshold, currently set at $150,000, and not just those focused on deployed activities or staff augmentation contracts. While Congress excluded research and development and military construction from the inventory in 2009, this exclusion did not result in better targeting of the direct labor hours collection requirement to meet congressional intent for the ICS.[10]

Difficulties with implementation of the ICS requirement resulted in persistent concerns regarding both the utility of the types of data collected in the ICS and the methods used for collecting these data, eventually leading House of Representatives conferees for the National Defense Authorization Act (NDAA) for FY 2016 to direct the Secretary of Defense to examine the approach the department was taking to comply with 10 U.S.C. 2330a(c). As part of this examination, the House conference report language directed the Secretary of Defense to determine whether DoD was producing a product that enhanced the oversight of service contracting activities and to submit a report to the congressional defense committees explaining the results of that examination and outlining efforts to better manage contractor and civilian personnel costs within DoD. The conferees further encouraged the Secretary of Defense to investigate and pursue existing DoD and service component IT systems that could present an alternative solution to the current ICS and that could be used in a timely manner to provide data relevant to strategic workforce planning.[11]

---

[10] Public Law 111-84, National Defense Authorization Act for Fiscal Year 2010, October 28, 2009.

[11] U.S. House of Representatives, *National Defense Authorization Act for Fiscal Year 2016*, conference report to accompany H.R. 1735, September 2015. Note that although the con-

Objectives of This Study

DoD sought RAND's assistance in responding to this congressional mandate. To support the Secretary of Defense in meeting the conferees' directive, we conducted this research with the following four objectives:

1. Contextualize the policy intent behind the congressional requirement for an ICS to determine the ultimate goals of collecting data on contracted services.
2. Define the types of data on contracted services that should be collected to ensure the greatest utility to DoD in the context of strategic planning and decisionmaking processes and outcomes.
3. Assess the methods currently used by DoD and other federal agencies to generate the ICS or other service contract inventories.
4. Identify alternative methods of collecting, processing, and reporting data on contracted services to meet the ultimate goals of the ICS and to facilitate DoD's strategic planning and decisionmaking efforts across manpower, budgeting, and acquisition functional communities, as appropriate.

This report fulfills these objectives by answering the following research questions:

- What is the congressional intent underlying the ICS requirement?
- To what extent does the ICS provide data on service contracts that are useful to DoD and congressional stakeholders?
- To the extent that there are gaps between the ICS data and data that would be most useful to DoD and congressional stakeholders, *why* do those gaps exist?

ference report language focuses on data relevant to strategic workforce planning, we were tasked with assessing systems and means of producing data relevant to DoD strategic planning more broadly, including budget and acquisition planning in addition to strategic workforce management.

- What relevant lessons could DoD learn from the experiences of non-DoD federal agencies for collecting, reporting, and using data on service contracts?
- How, if at all, could existing databases or information technology (IT) systems present a timely solution and provide data relevant to strategic workforce planning?
- What are the strengths and weaknesses of different methods for estimating and reporting on DoD's use of contractor personnel?

Study Approach

To answer these research questions, we employed a multifaceted approach incorporating a review of relevant legislation and literature, interviews with numerous stakeholders, and analyses of relevant data on service contracts. The aim of the review of relevant statutes, policies, and economic and historical literature was to understand (1) the statutorily required elements of an ICS; (2) the underlying intent driving the requirement for an ICS, as well as the intent driving similar requirements pertaining to data collection on service contracts by other federal agencies; and (3) the potential for particular contracted work to be converted to performance by federal civilian employees or military personnel, along with the economic concepts relevant to such considerations.

With the ultimate goals of improving service contract data collection efforts across DoD in mind, we conducted interviews with key stakeholders in DoD to identify data elements that stakeholders do find—or would find—most useful to meet the ultimate aims of the congressional requirement for an ICS, as well as the feasibility and effort entailed in collecting them.[12] Using a semistructured interview approach, we asked DoD interviewees a selection of questions

[12] These interviews are attributed anonymously throughout this report in compliance with the U.S. Federal Policy for the Protection of Human Subjects (also known as the Common Rule). Organizational affiliation is included in the citation for each anonymous interviewee to give a sense of background and experience, but it should be noted that interviewees were not asked to represent their organizations in a confidential way. While interviewees were

that depended on their position and level of experience with the ICS and related data systems. Questions ranged from tactical-level queries (about interviewees' experiences with the ICS, the time necessary to compile the data for the ICS report, and the number of staff necessary to work on ICS-related data reporting) to higher-level strategic questions, such as, "What are the main goals of data collection on service contracts, from your office's perspective, and how would you ideally use such data?" Other questions focused more specifically on the Contractor Manpower Reporting Application (CMRA) system used for collecting contractor direct labor hours for ICS reporting, asking about its ease of use and the extent to which the system added value to existing databases on service contracts. We also asked DoD interviewees whether they had verified ICS data quality at any point or knew of any process to do so, as well as questions about how they ensured that service contractors were not performing inherently governmental functions and the extent to which the ICS data were useful for that purpose. We reference the information gleaned from these confidential interviews later in this report, when we discuss our gap analysis highlighting where the ICS falls short of meeting its ultimate intent or requires an excessive amount of data collection.

Similarly, we interviewed congressional staffers involved in the development of the ICS requirement, DoD service contractors who have been affected by the ICS requirement, and subject-matter experts from policy research organizations. Interviews with congressional staffers focused on the ultimate congressional aims for the ICS, as well as the staffers' satisfaction with DoD actions and reporting to date. Interviews with service contractors centered around their experiences collecting and submitting data to fulfill ICS requirements, and interviews with subject-matter experts were intended to increase our understanding of the history of the ICS and key issues related to the system as currently implemented.

Finally, we interviewed representatives from non-DoD federal agencies responsible for pulling together data for their own analo-

---

asked to respond based on their professional experiences, they were in all cases speaking for themselves rather than for their organizations in an official capacity.

gous service contract inventory requirements. These interviews were intended to highlight best practices or lessons learned from these agencies' experiences that might be applicable to DoD's ICS requirement. Table 1.1 shows the range of organizations with which interviewees were affiliated at the time of our study. In total, we conducted interviews with 83 individuals across the organizations of interest. Detailed interview protocols for each population interviewed for this study can be found in Appendix D.

Third, we analyzed relevant data in the ICS, the Federal Procurement Data System—Next Generation (FPDS-NG), and National Defense Budget Estimates to identify trends in DoD spending on services over time and to forecast future spending. The objective of this examination was to develop samples of potential analyses that Congress might find useful to meet the underlying goals of the ICS requirement and that rely on existing data sources that do not require the same expenditure of resources and time needed to collect data on direct labor hours through CMRA.

This report combines the findings of the data analysis, the gap analysis derived from stakeholder interviews, and outputs from the literature review to develop recommendations regarding the types of data that could be collected cost-effectively to meet the congressional intent underlying the ICS. Finally, the report documents an economic analysis to develop alternative methods for estimating contractor personnel data to facilitate DoD's strategic planning and decisionmaking efforts, and it assesses the relative strengths and weaknesses of each proposed alternative.

Organization of This Report

The remainder of the report is organized as follows. In Chapter Two, we review relevant policies, statutes, and other literature and discuss the congressional intent underlying the ICS requirement (as indicated by both statutory guidance and interviews with congressional staff). Chapter Three draws on findings from our interviews with congressional staff, DoD stakeholders, non-DoD federal agency officials, and

Table 1.1
Organizations Represented in Interviews

Organization	Office
Office of the Secretary of Defense	Office of the Assistant Secretary of Defense for Manpower and Reserve Affairs
	Office of the Under Secretary of Defense for Acquisition, Technology, and Logistics
	Office of the Under Secretary of Defense for Personnel and Readiness
	Office of the Under Secretary of Defense, Comptroller
	Office of the Deputy Assistant Secretary of Defense for Maintenance, Policy, and Programs
	Office of the Deputy Assistant Secretary of Defense for Civilian Personnel Policy
	Defense Procurement and Acquisition Policy
Military services	Office of the Assistant Secretary of the Army for Manpower and Reserve Affairs
	Army Program Analysis and Evaluation
	Office of the Assistant Secretary of the Navy for Manpower and Reserve Affairs
	Office of the Assistant Secretary of the Navy for Research, Development, and Acquisition
	Office of the Assistant Secretary of the Air Force for Manpower and Reserve Affairs
	Office of the Assistant Secretary of the Air Force for Acquisition
	Office of the Air Force Deputy Chief of Staff for Manpower, Personnel, and Services (A1)
	Air Force Program Executive Office
Other defense entities	Defense Human Resources Agency
	Defense Manpower Data Center
	Defense Acquisition University

Table 1.1—Continued

Organization	Office
Congressional staff	House Armed Services Committee Staff
	Senate Armed Services Committee Staff
	Senate Homeland Security and Governmental Affairs Committee Staff
	Other professional staff, as appropriate
Policy research organizations	U.S. Government Accountability Office
Non-DoD federal agencies	U.S. Department of the Treasury
	U.S. Department of Homeland Security
	U.S. Department of Education
	Office of Management and Budget
Contractors	Various service contractors

a sample of service contractors to identify the strengths and weaknesses of the substance and process of the ICS as currently conceived, as well as of the CMRA IT system used for collecting ICS-relevant data. Chapter Four then explores the variation in types of DoD service contracts, underscoring the difficulty of devising a strong analytical product housing a large-scale inventory of such contracts that is also useful as a basis for decisionmaking. Chapter Five examines the prospects for utilizing existing databases and systems to provide data and analysis able to meet congressional goals for the ICS, providing numerous sample analyses of FPDS-NG data. Chapter Six then weighs the risks and benefits of the ICS as currently conceived (including the use of CMRA for collecting data on direct labor hours from contractors) against the pros and cons associated with alternative methods for meeting the congressional intent of the ICS requirement. Chapter Seven concludes the report with a discussion of our research findings and recommendations to facilitate the collection of data on contracted services that will have utility in meeting the ultimate aims of the ICS across DoD.

CHAPTER TWO

Review of the Relevant Literature and Congressional Intent Underlying the ICS Requirement

How the U.S. government procures goods and services is a persistent concern in Congress, with the drivers of this concern varying over time. The federal government's long-standing policy has been that it "will not start or carry on any commercial activity to provide a service or product for its own use if such product or service can be procured from private enterprise through ordinary business channels."[1]

Interest in contracting or outsourcing goods and services—that is, "the transfer of a support function traditionally performed by an in-house organization to an outside service provider"—has generally intensified over time, with some exceptions.[2] Meanwhile, the interest of private firms in focusing on "core competencies" and contracting out for others has similarly intensified.[3] Continuing IT developments have reduced costs between buyers and suppliers, allowing governments to contract out the production of many goods and services to an increasing extent.

Interest in government contracting for goods and services particularly increased in the United States under the Reagan administration and in the United Kingdom under the Thatcher administration in the 1980s, with both governments seeking to increase efficiency

---

1 Executive Office of the President, Bureau of the Budget, "Commercial-Industrial Activities of the Government Providing Products or Services for Governmental Use," Washington, D.C., Bulletin No. 55-4, January 15, 1955.

2 Office of the Under Secretary of Defense for Acquisition and Technology, 1996, p. 7A.

3 Gary Hamel, and C. K. Prahalad, *Competing for the Future*, Boston, Mass.: Harvard Business School Press, 1994.

through contracting. Interest in contracting out for goods and services increased further after the Cold War and during the 1991 Gulf War, when then–Secretary of Defense Dick Cheney sought to boost the number of deployed contractors and otherwise privatize military functions.[4] At the same time, concerns about the size and capabilities of the acquisition workforce, contracting techniques and approaches, service acquisitions, and operational contract support led the U.S. General Accounting Office (now the U.S. Government Accountability Office, or GAO) to identify DoD contract management as being at high risk for fraud, waste, or mismanagement.[5]

In the face of decreasing defense budgets in the 1990s, a Defense Science Board Task Force on Outsourcing and Privatization suggested that DoD "could generate savings of up to $7 to $12 billion annually" by FY 2002 through aggressive contracting measures.[6] Specifically, it suggested DoD consider all non–combat support services except "inherently governmental" services and those "for which no adequate private sector capability exists or can be expected to be established" as potential candidates for outsourcing.[7] A subsequent GAO study suggested that the likely level of savings was only one-third what the Defense Science Board Task Force suggested but concurred that savings were possible and, indeed, that DoD had already achieved more than $700 million in savings through contracting and similar initiatives.[8]

The Federal Activities Inventory Reform Act of 1998 attempted to encourage privatization of government activities that were "not inherently governmental functions" by "set[ting] forth procedures for determining whether such activities should be performed under contract

---

[4] Molly Dunigan, *Victory for Hire: Private Security Companies' Impact on Military Effectiveness*, Stanford, Calif.: Stanford University Press, 2011.

[5] U.S. Government Accountability Office, *High-Risk Series: An Update*, Washington, D.C., GAO-15-290, February 2015a.

[6] Office of the Under Secretary of Defense for Acquisition and Technology, 1996, p. 1A.

[7] Office of the Under Secretary of Defense for Acquisition and Technology, 1996, p. 53.

[8] U.S. General Accounting Office, *Outsourcing DoD Logistics: Savings Achievable but Defense Science Board's Projections Are Overstated*, Washington, D.C., NSIAD-98-48, December 1997.

with private sector sources or in-house using governmental facilities and personnel."[9] The legislation required agencies to submit to OMB a list of activities that were not inherently governmental but were performed by federal employees for subsequent review and possible competition. At the time of passage, one legislator estimated that 1.4 million federal employees performed activities that were commercial in nature, while another claimed that four departments in addition to DoD had at least 10 percent of their full-time equivalent (FTE) employees performing commercial activities.[10]

Privatization of defense functions in the United States continued into George W. Bush's administration. To support management of contracted services, Congress passed the NDAA for FY 2002, codified at 10 U.S.C. 2330a, requiring the Secretary of Defense to "establish and implement a management structure for the procurement of services" by DoD.[11] The legislation also directed DoD to conduct spending analyses of its purchases for contract services. This legislation was prompted, in part, by congressional concern that DoD had "never conducted a comprehensive spending analysis of its service contracts and has made little effort to leverage its buying power, improve the performance of its service contractors, rationalize its supplier base, or otherwise ensure that its dollars are well spent."[12]

Operations in Afghanistan and Iraq after the September 2001 terrorist attacks led to extensive, sudden growth of use of operational contract support services—and, ultimately, concerns regarding oversight, contractor accountability, and waste, fraud, and abuse. As defense budgets once again grew, DoD doubled the volume of goods and services it was purchasing from the private sector.[13] Similarly, DoD obligations

---

9 U.S. Senate, Committee on Governmental Affairs, *Federal Activities Inventory Reform Act of 1998*, Report 105-269 to accompany S. 314, July 28, 1998.

10 U.S. Senate, Committee on Governmental Affairs, 1998.

11 Public Law 107-107, National Defense Authorization Act for Fiscal Year 2002, December 28, 2001.

12 U.S. Senate, Committee on Armed Services, *National Defense Authorization Act for Fiscal Year 2002*, Report 107-62 to accompany S. 1416, September 12, 2001, p. 326.

13 Josh Rogin, "The Hidden Price of a Buying Spree," *CQ Weekly*, July 23, 2007.

for service contracts (not including research, development, testing, and evaluation [RDT&E]) increased from $53 billion in FY 2000 to $104 billion in FY 2006.[14]

A reduction by half of the Defense Contract Management Agency civilian staff since 1993 made managing this large volume of goods and services even more difficult.[15] The Commission on Wartime Contracting in Iraq and Afghanistan estimated that between $31 billion and $60 billion was lost in contract waste and fraud in Iraq and Afghanistan because of poorly planned or executed contracts and a federal lack of capacity to monitor contracts.[16]

Congress expressed particular concerns about general categories used to justify budgets for specific services. It claimed that budget materials submitted by the Secretary of Defense "identified categories of information that . . . should be specifically identified under the 'other purchases' category in the summary of price and program growth."[17] Accordingly, it directed the "the Secretary to identify in budget justification material . . . the costs for outsourcing and privatization, information technology contracts, other base support contractual services, other training contractual support, and military personnel contract support."[18] Congress mandated the elements of such an inventory of contracted services in the NDAA for FY 2008.

The ICS Requirement and Congressional Goals for Its Use

The inventory mandated by the FY 2008 NDAA was to include contractor functions and missions, contracting organization, funding

---

[14] U.S. Government Accountability Office, *Contract Management: DOD Vulnerabilities to Contracting Fraud, Waste, and Abuse*, Washington, D.C., GAO-06-838R, July 2006.

[15] Rogin, 2007.

[16] Commission on Wartime Contracting in Iraq and Afghanistan, 2011.

[17] U.S. House of Representatives, Committee on Armed Services, *National Defense Authorization Act for Fiscal Year 2006*, Report 109-89 to accompany H.R. 1815, May 20, 2005, p. 302.

[18] U.S. House of Representatives, Committee on Armed Services, 2005, p. 302.

source for the contract, fiscal year in which contracted activity began, number of full-time contractor employees or equivalent paid for the contracted activity, and a determination of whether it was a personal services contract (i.e., a contract creating an employer-employee relationship between the government and the contractor—a specific subset of what is referred to throughout this report as a *staff augmentation* contract).[19] DoD was also directed to publish the inventory and to use it to identify contracts that were providing personal services or inherently governmental services, as well as to address congressional concerns about managing the service supply base.

Congress also made similar requirements of civilian agencies. Citing concern over contracts with poor performance, excessive costs, or inferior quality, it directed OMB to "establish a pilot program to develop and implement an inventory to track the cost and size (in contractor manpower equivalents) of service contracts."[20] The legislation directed OMB to establish the pilot program in at least three Cabinet-level departments, including one that contracted annually for more than $10 billion in services, one that contracted annually for between $5 billion and $9 billion in services, and one that contracted annually for less than $5 billion in services.

Subsequent legislation extending the inventory requirement to other civilian agencies requested elements similar to those in the DoD ICS: description of services purchased, organizational component administering the contract, total dollars obligated and invoiced for the services, contract type and date, contractor name and address, number and location of contractor and subcontractor FTEs for direct labor, and whether it was a personal services contract or awarded noncompetitively.[21] Legislators who supported these provisions noted the lack of information regarding reliance on service contractors and whether agencies "have the right balance of contractor and in-house

---

[19] Public Law 110-181, National Defense Authorization Act for Fiscal Year 2008, January 28, 2008.

[20] Public Law 110-161, Consolidated Appropriations Act, 2008, December 26, 2007.

[21] Public Law 111-117, Consolidated Appropriations Act, 2010, December 16, 2009.

resources needed to accomplish their missions."[22] They also required a review of the inventory data to ensure that personal services contracts had been entered appropriately and that contractor personnel were not performing inherently governmental functions.

Congress similarly had concerns that the Intelligence Community relied too heavily on contractors and could not account for their annual number or costs.[23] Such concerns prompted the Intelligence Community's chief human capital officer to compile an annual core contract personnel inventory in FY 2007. Congress codified these requirements in 2010, directing the Office of the Director of National Intelligence to compile an annual report on personal services contracts, including their costs, number of personnel, compensation relative to that of government personnel performing similar work, and plans for conversion to government positions.[24] GAO, however, would find that the data had limited comparability due to their varying definition of core contract personnel and other shortcomings.[25]

In FY 2011, Congress recommended that DoD use the ICS not just to track acquisitions but also to facilitate "human capital planning" and efforts to ensure that DoD was using "the right mix of military personnel, civilian employees, and contractors" and to clarify when contracting for services was appropriate.[26] The House Committee on Appropriations also noted the context of such actions within broader government-wide efforts to "strengthen oversight, end unnec-

---

[22] U.S. House of Representatives, *Departments of Transportation and Housing and Urban Development, and Related Agencies Appropriations Act, 2010*, conference report to accompany H.R. 3288, December 8, 2009.

[23] The Central Intelligence Agency, the Office of the Director of National Intelligence, and intelligence components within the departments of Energy, Homeland Security, Justice, State, and the Treasury. See U.S. Government Accountability Office, *Civilian Intelligence Community: Additional Actions Needed to Improve Reporting on and Planning for the Use of Contract Personnel*, Washington, D.C., GAO-14-204, January 2014a.

[24] Public Law 111-259, Intelligence Authorization Act for Fiscal Year 2010, October 7, 2010.

[25] U.S. Government Accountability Office, 2014a.

[26] U.S. House of Representatives, Committee on Armed Services, *National Defense Authorization Act for Fiscal Year 2011*, Report 111-491 to accompany H.R. 5136, May 21, 2010, p. 271.

essary no-bid and cost-plus contracts, [and] maximize the use of competitive procurement processes."[27]

The multiple drivers underlying the ICS requirement mandated in the FY 2008 NDAA—a desire for better cost control over service contracts, more refined budgetary planning with regard to service contracts, and, relatedly, informing and facilitating strategic workforce planning and potential insourcing decisions—were echoed in interviews with congressional staff members involved in devising the ICS requirement and responsible for overseeing its implementation. We interviewed 11 House and Senate staffers, both Democrats and Republicans, over the course of this study. They exhibited some disagreement about the original motivations for the inventory, but they were mostly aligned in terms of how they explained what Congress wanted the DoD to do with the ICS. Seven of our interviewees emphasized that DoD should use the inventory to inform both budgeting and strategic workforce planning. [28] Only one interviewee mentioned that the data should be used to ensure that the department is getting "the value for the money."[29] Another interviewee explained how the different congressional drivers of the inventory were at odds with each other.[30] In addition, five interviewees identified congressional oversight of DoD activities as another important aspect of the ICS.[31] All the interviewees indicated disappointment with DoD's actions and deliverables with respect to the inventory.

One staffer thought that the ICS should be used to inform all levels of DoD decisionmaking, in terms of both strategic workforce planning and budgeting.[32] Another mentioned that Congress wanted the department to use this information for strategic planning pur-

---

[27] U.S. House of Representatives, Committee on Appropriations, *Department of Defense Appropriations Bill, 2010*, Report 111-230 to accompany H.R. 3326, July 24, 2009, p. 6.

[28] Interviews with congressional staff, January–February, 2016.

[29] Interview with a congressional staffer, January 2016.

[30] Interview with a former congressional staffer, January 2016.

[31] Interviews with congressional staff, January–February, 2016.

[32] Interview with a congressional staffer, January 2016.

poses and therefore wanted actual direct labor hours, not estimates.[33] A third congressional staff interviewee indicated a desire for DoD to just be "able to answer the question, by service and defense agencies, how many service contracts do you have, [and] how much is it costing you?"[34]

Less evident in the ICS statutory language prior to 2013 was the aim of increasing congressional oversight of *deployed* operational contract support activities, and particularly the number of contractors operating under DoD contracts in operational theaters.[35] Yet several interviewees recognized this as a primary driver of the ICS requirement, noting that the impetus for the ICS requirement sprung from concern over DoD contractor activities early in Operation Iraqi Freedom. One interviewee explained that

> during the war in Iraq, when services contracting went through the ceiling in terms of expenditures, and issues with security firms arose. . . . The committee wanted visibility on what we're spending money on, where we're spending it, and what kinds of functions are being performed.[36]

The interviewee explained that the ICS was a "direct outgrowth of security contractor issues and [related] well-publicized events."[37] The same interviewee also indicated that Congress wanted data on contrac-

---

[33] Interview with a congressional staffer, January 2016.

[34] Interview with a congressional staffer, January 2016.

[35] The FY 2013 NDAA (Public Law 112-239, January 2, 2013) required DoD, the U.S. Department of State, and the U.S. Agency for International Development to collect data, including the total number of contractor personnel, on contract support for future contingencies outside the United States involving combat operations.

[36] Interview with a congressional staffer, January 2016.

[37] Interview with a congressional staffer, January 2016. Although this interviewee did not refer to them by name, high-profile incidents involving security contractors in Operation Iraqi Freedom included the well-publicized murder and hanging of the charred remains of four Blackwater contractors from a bridge in Fallujah in 2006 and the 2007 Nisour Square shootings in Baghdad, in which Blackwater contractors shot at unarmed Iraqi civilians, killing 17 and injuring 24 others. See, for example, Bill Sizemore and Joanne Kimberlin, "Blackwater, Part 4: When Things Go Wrong," *Virginian-Pilot*, July 26, 2006, and James

tor FTEs because it wanted visibility into how many contractors DoD had, in part to gain greater oversight over operational contract support activities. Another congressional staffer mentioned that understanding the workforce mix in theater is very important,[38] and a third stated that knowing the number of personnel on a contract is very important for workforce mix decisions, budgeting, *and* operational questions in theater.[39]

Yet another congressional interviewee highlighted an inherent contradiction in the congressional requirement for the ICS, explaining that there were two congressional drivers of the ICS requirement, each specifying different data collection requirements. One driver of the ICS was to devise a product that would enable DoD to conduct spend analyses to better control service contract costs. The second driver motivating Congress to mandate the development of an ICS was concern regarding the use of service contracts for staff augmentation purposes and a related desire to understand where functions performed under such staff augmentation contracts could and should be insourced to better control costs while strategically managing the defense workforce.

This interviewee explained, "The purpose of an inventory which looks at contracted services in comparison to civilian employees is different from an inventory that compares contracted services to acquisition of products."[40] If you are looking to compare civilian and contractor workforces, you are only interested in the subset of contracted services for which contractors are directly comparable to civilian FTEs. This type of inventory would collect contractor FTEs, but for a select group of service contracts only—that is, those contracts under which only labor is outsourced, with the purpose of augmenting government personnel capabilities. "If you focused the universe in the right way, you could do meaningful comparisons" of costs and trade-offs and

---

Glanz and Alissa J. Rubin, "Blackwater Shootings 'Murder,' Iraq Says," *New York Times*, October 8, 2007b.

[38] Interview with a congressional staffer, January 2016.

[39] Interview with a congressional staffer, February 2016.

[40] Interview with a former congressional staffer, January 2016.

analyze whether contractors are performing inherently governmental functions, the interviewee said.[41] The purpose of an inventory that compares product acquisition to service spending, on the other hand, would allow a spend analysis, providing a better understanding of what is being purchased, from whom it is being purchased, what contract type is best, and whether there is the right number of contracts for a given service. Counting contractor FTEs does not make sense under this approach.[42] In sum, this interviewee emphasized that congressional goals for the ICS required "two different types of analysis for two types of universes," and, as a result, the ICS as currently conceived is "a product that is not as useful as it could be."[43] In Chapter Four, we return to this discussion of different goals for the ICS, requirements for different types of data collection, and the need to distinguish between different types of service contracts.

The Current Status of the ICS and Related DoD Activities

Today, the inventory is intended to serve a variety of budgetary and planning needs. As originally mandated, it is to provide information to support submission of budgetary information. The NDAA for FY 2012 stipulated that DoD was to use it to ensure that the total amount obligated for contract services was not increasing.[44] The NDAA for FY 2013 stipulated that DoD was to use it to achieve savings in total funding of government civilian and contractor workforces matching those achieved for basic military personnel pay resulting from reductions in end strength.[45] The inventory was also to be used to create a biennial strategic workforce plan, to ensure that contracts were

[41] Interview with a former congressional staffer, January 2016.

[42] Interview with a former congressional staffer, January 2016.

[43] Interview with a former congressional staffer, January 2016.

[44] Public Law 112-81, National Defense Authorization Act for Fiscal Year 2012, December 31, 2011.

[45] Public Law 112-239, 2013.

not awarded for personal or inherently governmental services, and to help identify categories of functions for insourcing.

Broadly speaking, DoD has taken action over time to address congressional concerns about contract services from a personnel management perspective. In 2005, it updated and outlined criteria for managing its mix of military, civilian, and contractor personnel.[46] Its 2006 *Quadrennial Defense Review Report* renewed emphasis on managing the total force, comprising all military, civilian, and contractor personnel.[47] DoD further sought to shift resources from contract support to more than 33,000 new civilian manpower authorizations from 2010 to 2014, which would effectively insource some contracted services—that is, move work from contractors to civilian or military positions within the government.[48]

In response to the ICS requirement, DoD issued an instruction on the policy and procedures for determining the appropriate mix of military and civilian manpower and private-sector support.[49] It identified potential costs associated with manpower to be considered in decisionmaking processes. It also updated instructions establishing policies, delineated responsibilities, provided direction for the acquisition of contracted services, and established and implemented a management structure for the acquisition of services.[50]

Moreover, DoD has taken steps to document numbers of contractor FTEs and to launch an enterprise-wide system, eCMRA, based on

---

[46] U.S. Department of Defense Directive 1100.4, *Guidance for Manpower Management*, February 12, 2005.

[47] U.S. Department of Defense, *Quadrennial Defense Review Report*, Washington, D.C., February 6, 2006b.

[48] U.S. Department of Defense, *Resource Management Decision 802*, Washington, D.C., April 8, 2009; Office of the Deputy Secretary of Defense, "Insourcing Contracted Services: Implementation Guidance," memorandum, May 28, 2009.

[49] U.S. Department of Defense Instruction 1100.22, *Policy and Procedures for Determining Workforce Mix*, April 12, 2010.

[50] U.S. Department of Defense Instruction 5000.74, *Defense Acquisition of Services*, January 5, 2016.

the Army-developed CMRA and as directed by Congress.[51] As part of its long-term plan to ensure DoD-wide compliance with 10 U.S.C. 2330a, the Office of the Under Secretary of Defense for Acquisition, Technology, and Logistics and the Office of the Under Secretary of Defense for Personnel and Readiness announced in late 2011 a plan to have the Office of the Deputy Chief Management Officer, as DoD's lead business systems integrator, manage a working group of stakeholders to support the establishment of a common hardware and software support platform to meet the requirements of 10 U.S.C. 2330a.[52] The working group's aim was to expedite the analysis of how best to implement the statutory requirement leveraging existing systems, such as the Army's CMRA system.[53] These efforts led then–Secretary of Defense Leon Panetta to commit to the Army CMRA solution in December 2011 and outline its enterprise-wide implementation by the end of FY 2012.[54] This implementation plan called for the establishment of the Total Force Management Support Office under the auspices of the Office of the Under Secretary of Defense for Personnel and Readi-

[51] See, for example, Office of the Secretary of Defense, "Enterprise-Wide Contractor Manpower Reporting Application," memorandum, November 28, 2012, outlining joint guidance from the Office of the Under Secretary of Defense for Acquisition, Technology, and Logistics and the Office of the Under Secretary of Defense for Personnel and Readiness on eCMRA. See also Public Law 112-10, Department of Defense and Full-Year Continuing Appropriations Act of 2011, April 5, 2011, which instructs DoD to adopt the Army's CMRA system department-wide.

[52] See Appendix F of this report for a list of references citing defense components' plans for compliance with Section 8108(c) of the Department of Defense and Full Year Continuing Appropriations Act of 2011 (Public Law 112-10, 2011), which required the secretaries of the Army, Navy, and Air Force and the directors of the defense agencies and field activities (in coordination with the Office of the Under Secretary of Defense for Personnel and Readiness) to report to the congressional defense committees within 60 days of the act's enactment their plan for documenting the number of full-time contractor employees (or equivalent) they employed, as required by 10 U.S.C. 2330a.

[53] Office of the Secretary of Defense, memorandum on DoD's compliance plan for Public Law 112-10, Section 8108(c), Enclosure 1, November 22, 2011; interviews with officials in the Office of the Secretary of Defense (OSD), January 13 and 15, 2016; interview with a former OSD official, January 28, 2016.

[54] Secretary of Defense Leon Panetta, letter to Howard P. "Buck" McKeon, chair of the House Armed Services Committee, December 20, 2011.

ness to support the use of eCMRA and the process for collecting and reporting ICS data. However, these efforts were put on hold following leadership changes in the department.[55]

As it currently stands, the ICS is produced about one year following the end of the fiscal year for which data are being reported, and it is captured in two publicly available formats: the Office of the Under Secretary of Defense for Acquisition, Technology, and Logistics ICS report to Congress, and the 37 different defense component spreadsheets on the Defense Procurement Acquisition Policy (DPAP) website. The ICS is produced using the CMRA system. The Army first developed the CMRA system, but now there are four separate "instances," or versions, of the system—one each for the Army, Air Force, and Navy and one for the other defense agencies. As currently planned, the different instances of CMRA will be combined into one "enterprise-wide" eCMRA within the next several years, and all instances are now being moved under Defense Manpower Data Center stewardship.

We were unable to access CMRA data for this study, as access is limited to protect contractors' proprietary data from competitors. However, it is critical to note that even without access to restricted CMRA data, we were able to use the publicly available ICS data published on the DPAP website (which reports contract number and information on direct labor hours) to link ICS-reported direct labor hours to particular service contractors using contract number information publicly available on the FPDS-NG website. We also found issues of completeness and lack of quality when we analyzed the ICS data and compared them with FPDS-NG data. These issues are explored in greater detail in Appendix C.

All in all, concerns remain over how well DoD is complying with mandates regarding the inventory of contracted services. These concerns led Congress to "direct the Secretary of Defense to examine the approach the Department is taking to comply with" provisions for the inventory, "and determine whether it is or is not producing a product

---

[55] Project on Government Oversight, letter to Secretary of Defense Chuck Hagel, November 25, 2014; interviews with OSD officials, January 13 and 15, 2016; interview with a former OSD official, January 28, 2016.

that enhances the oversight of service contracting activities," submitting a report on that effort. As noted in Chapter One, this report documents research conducted in response to that requirement.

In the next chapter, we explore how well the ICS has met—or can meet—congressional objectives and DoD needs. To summarize, congressional goals for the ICS are to enhance cost control and budgetary planning related to service contracts, to enable strategic workforce planning that incorporates consideration of service contractors, and, as discussed in our interviews, to maintain congressional oversight of the numbers and activities of deployed operational support contractors in theater. In some cases, as we discuss, other sources of data—though sometimes dismissed by Congress and others—may provide more timely insights to fulfill some of these objectives, while more carefully tailoring aspects of the ICS requirement to specific types of contracts and developing a plan for processing and analyzing ICS data may help address others.

CHAPTER THREE

Assessing the Success of the ICS in Meeting Congressional Objectives and DoD Needs

As noted in Chapter Two, congressional goals for the ICS are multifaceted: to control costs and inform budgetary decisionmaking, to allow for more informed strategic workforce planning and insourcing decisions, and to provide Congress with greater oversight of operational contract support activities in theater. In an effort to determine the extent to which the ICS is meeting the needs of Congress and DoD, we conducted semistructured interviews with 11 congressional staff members and 41 DoD officials. In an effort to gain a broader understanding of the ease or burden of ICS reporting, as well as to derive best practices potentially applicable to DoD, we also interviewed ten non-DoD federal officials, 19 individuals working for nine service contracting firms that provide services to DoD, and two subject-matter experts from a policy research organization, for a total of 83 interviewees.

We categorized our interviewees by relevant functional community and analyzed their comments in terms of views on the types of data most desired, gaps between the ICS data and the data interviewees wanted, and specific views on the utility of collecting data on direct labor hours. It became apparent that the ICS, created using a combination of both contractor-reported and component-estimated direct labor hours from the military services' different versions of the CMRA, is falling short of meeting the needs of both Congress and DoD. While most congressional staff interviewed for the study were in agreement regarding the failings of the ICS, DoD officials offered a wide spectrum of views on its overall utility, the utility of the unique direct labor hour data captured in CMRA for reporting in the ICS, and the data that would be most relevant for decisionmaking related to service contracts.

The ICS Falls Short of Meeting Congressional Objectives

It was widely recognized that the initial inventory suffered from several shortcomings, as elaborated in numerous GAO reports. While the military services identified nearly $100 billion in contract services and nearly 600,000 contractor FTE workers providing these services, they used different methods and data systems to do so, relying in some cases on contractor-entered data and raising concerns about accuracy and thoroughness.[1] Inventory data were more consistent for FY 2009 but still reflected differences in reporting methods among the military services and were not comparable with data from FY 2008.[2]

Congress would seek to clarify and strengthen the provisions of the inventory in response to identified shortcomings. The NDAA for FY 2011 sought to ensure that the information on contractor labor hours were actual, direct labor hours, rather than estimates from other sources, and gave responsibility for developing the inventory to the Office of the Under Secretary of Defense for Personnel and Readiness.[3]

While it was hoped that centralizing the inventory data collection would help improve consistency, use of differing sources continued to lead to some concerns. For the FY 2010 inventory, DoD relied primarily on FPDS-NG data, which do not identify more than one type of service for each contract action or include the number of contrac-

---

[1] U.S. Government Accountability Office, *Defense Acquisitions: Observations on the Department of Defense Service Contract Inventories for Fiscal Year 2008*, Washington, D.C., GAO-10-350R, January 2010a.

[2] U.S. Government Accountability Office, *Defense Acquisitions: Further Action Needed to Better Implement Requirements for Conducting Inventory of Service Contract Activities*, Washington, D.C., GAO-11-192, January 2011.

[3] Public Law 111-383, National Defense Authorization Act for Fiscal Year 2011, January 7, 2011.

tor personnel.[4] In contrast with FPDS-NG data, the CMRA data collected by the Army complied with all legislative requirements for the inventory and included contractor-reported data on direct labor hours, function, and mission at the contract line-item level. Congress subsequently required the Air Force and Navy to modify the Army's CMRA system for use in their inventories.[5]

The format in which the ICS data are reported to Congress also remains a source of concern, with regard to the data's utility for congressional oversight. Indeed, as noted in Chapter Two, many congressional staff interviewed for this study felt that the format in which ICS data are reported was not useful and hindered congressional assessment. One congressional staffer did not see the value in extensively detailed information, explaining that Congress does not have the bandwidth to analyze data on "eaches." This interviewee's vision of the ICS as submitted to Congress was "five pages that show, by agency, how much money you're spending on contracts, what's the equivalent labor . . . and what that would mean for a civilian equivalent."[6] Another staff member echoed this point: "Congress doesn't need all the extraneous data being provided now. They need useful data."[7] Yet another congressional staffer expressed the opinion that Congress did not really understand what it was asking for and that "the right requirement for Congress would be a synthesized report at a higher level" that legislators could then use as the basis for further queries.[8] The interviewee noted that the statutory language was too detailed for what an oversight committee would need. "It's data," the interviewee pointed out, "not information."[9] It appears that Congress seeks *analysis*—that is, information and metrics—and not raw data from DoD on the topics rel-

---

[4] U.S. Government Accountability Office, *Defense Acquisitions: Further Actions Needed to Improve Accountability for DOD's Inventory of Contracted Services*, Washington, D.C., GAO-12-257, April 2012.

[5] Public Law 112-10, 2011.

[6] Interview with a congressional staffer, January 2016.

[7] Interview with a congressional staffer, January 2016.

[8] Interview with a congressional staffer, January 2016.

[9] Interview with a congressional staffer, January 2016.

evant to the ICS. However, this is not clearly specified in statute, which specifies the following congressional goals for the use of ICS products:

- Strengthen oversight.
- Reduce no-bid and cost-plus contracts.
- Maximize the use of competitive procurement processes.
- Conduct spending analyses of contract service purchases.
- Improve the performance of service contractors.
- Leverage buying power.
- Rationalize the supplier base.
- Track services provided in joint capability areas and assess whether the mix is appropriate for current operations and 10–15 years into the future.
- Inform insourcing decisions and determine the appropriate mix of personnel.
- Conduct total workforce planning that is fully integrated into the programming and budget processes.

Unfortunately, the ICS products do not fulfill the majority of these goals. DoD's FY 2014 ICS report to Congress in its aggregated format does not meet any of these goals, and the 37 defense components' ICS spreadsheets of all actions/invoices by contract number with reported or estimated FTEs (as published on the DPAP website) only partially meet one of these goals: to inform insourcing decisions and determine the appropriate mix of personnel.[10]

The ICS Similarly Falls Short of Meeting DoD Needs

Like Congress, stakeholders across DoD have different goals for data collection, making it difficult to determine the types of data that should

[10] However, without information on the relative extent to which various contractors utilize different inputs in addition to labor (e.g., technology, capital) when they produce a specific service, information on contractor FTEs cannot be used to make effective labor and productivity comparisons to accurately inform insourcing and workforce management decisions. This issue is discussed in greater detail in Chapter Four.

be collected and why. However, it is critical to understand the goals of data collection to know the relevant data points to collect. In this case, the lack of a unified departmental vision for the use of the ICS complicates the discussion regarding what the ICS should consist of and how best to collect and analyze the relevant data.

Views on the ICS exemplify bureaucratic politics in the most classic sense; that is, where you sit (in an office) defines where you stand (on the issue of the ICS).[11] This makes sense, as different offices in DoD play different roles in service contracting activities and thus have different perspectives on service contract data. Three DoD functional communities, in particular, are relevant to the ICS: the manpower and personnel community (represented primarily by officials in the Office of the Under Secretary of Defense for Personnel and Readiness), the budgeting community (primarily represented by officials in the Office of the Under Secretary of Defense, Comptroller), and the acquisition community (primarily represented by officials in the Office of the Under Secretary of Defense for Acquisition, Technology, and Logistics). We categorized the majority of DoD interviewees into one of these three functional communities, based on their organizational affiliations and feedback received during the interview about their current position and career background.

The Types of Data Most Desired by DoD Stakeholders

The manpower and personnel community in DoD is primarily interested in data that can inform manpower and personnel planning, as well as strategic workforce planning aimed in particular at shaping the total force. Driven by this overarching goal, this functional community seeks service contract data that point to level of effort—specifically, contractor FTEs. We interviewed 15 individuals from the DoD manpower and personnel community for this research. One of these interviewees noted that the ICS and CMRA have the most value for the manpower and programming community, as they allow "like comparisons" of the organic workforce by providing "a level of effort

---

[11] On the concept of bureaucratic politics, see Graham Allison and Philip Zelikow, *Essence of Decision: Explaining the Cuban Missile Crisis*, 2nd ed., London: Pearson, 1999.

measurable and comparable for services and workload similar to military end strength or civilian FTEs that can gauge how much DoD relies on contractors for certain types of work in certain places." They also allow value judgments to be made on this basis.[12] Another interviewee from this community indicated a need for data that enable total force planning to "help you complete the picture of where your workforce is, where your efforts are."[13] Echoing this point, a third interviewee from this community noted, "I want to make better labor decisions among our components."[14]

The budgeting community, however, is primarily interested in data that can inform budget processes. As such, this community desires data on the total cost of service contracts and data at an appropriate level of analysis to enable it to integrate well with budget object–class considerations. One of our four interviewees from the budgeting community highlighted this point, noting that "the focus should be on how much we're spending, not how many contractors we're getting."[15]

Meanwhile, the acquisition community is mainly interested in data that can inform acquisition planning and procurement decisions. Like the budgeting community, this community seeks data on the total cost of service contracts, but also on level of performance. Thirteen individuals from this community were interviewed for this research. One interviewee specified that ideally, they would like to be able to utilize service contract data for market analysis and to help improve the writing of requirements.[16] Another interviewee from the acquisition community went further:

> We would be interested in data that would allow us to compare individual contracted efforts and see if there's some sort of causal relationship between some of the things we do regarding statements of work, labor categories, workforce mix on that contract,

---

[12] Interview with an OSD official, January 2016.

[13] Interview with a former OSD official, January 2016.

[14] Interview with an OSD official, January 2016.

[15] Interview with an OSD official, January 2016.

[16] Interview with a defense official, January 2016.

> those types of data sets, that would allow us to identify best practices, causal outcomes both positive and negative, and some sort of labor rate optimization for tasking going forward.

In the words of a third interviewee from the acquisition community, "You don't care about people working on [the contract]. You care about the work being done."[17]

Views on the ICS Collection of Direct Labor Hours

A particular point of contention among DoD interviewees was the value of direct labor hours. As noted in Chapter Two, 10 U.S.C. 2330a requires the Secretary of Defense to submit to Congress an annual inventory of the activities performed during the preceding fiscal year, pursuant to contracts for services (and pursuant to contracts for goods, to the extent that services are a significant component of performance as identified in a separate line item of a contract) for or on behalf of DoD. To fulfill this requirement, Section 2330a further mandates that

> The entry for an activity on an inventory under this subsection shall include: the functions and missions performed by the contractor; the contracting organization; the component of the Department of Defense administering the contract, and the organization whose requirements are being met through contractor performance of the function; the funding source for the contract under which the function is performed by appropriation and operating agency; the fiscal year for which the activity first appeared on an inventory under this section; *the number of contractor employees, expressed as full-time equivalents for direct labor, using direct labor hours and associated cost data collected from contractors (except that estimates may be used where such data is not available and cannot reasonably be made available in a timely manner for the purpose of the inventory)*; a determination whether the contract pursuant to which the activity is performed is a personal services contract; and a summary of the data required to be collected for the activity.[18]

---

[17] Interview with an OSD official, January 2016.

[18] Title 10 U.S.C. 2330a(c)(2); emphasis added.

The requirement to collect contractor direct labor hours is significant, as both direct labor hours and nonlabor direct costs are the only unique data fields captured in the ICS—and reported through CMRA—that are not found in alternative databases, such as FPDS-NG or the System for Award Management (SAM).[19] Yet, as the statute alludes, estimates are used in place of actual contractor-reported direct labor hours in some instances. Indeed, in the FY 2014 ICS, contractor-reported direct labor hours were available for only 38 percent of ICS dollars across all of DoD, indicating that the remaining 62 percent of direct labor hours were estimated based on Army algorithms. However, the degree to which estimates are used varies across the military services and defense components, with the Army claiming that 100 percent of ICS dollars are associated with contractor-

---

[19] SAM is designed to provide information about recipients of federal awards (e.g., contracts or grants). Data elements are listed in the *SAM Functional Data Dictionary* (see U.S. General Services Administration, SAM Functional Data Dictionary, version 7.0, Washington, D.C., July 17, 2014). Any entity—defined as a vendor, grantee, or government agency involved in intragovernmental transactions—that seeks an award or grant from the federal government must register in SAM, with few exceptions outlined in the Federal Acquisition Regulation. Entities submit and revise their data online. SAM entity data come from various sources:

- Dun & Bradstreet is the authoritative source for the entity's name and physical address that corresponds to the Data Universal Numbering System number provided upon initial registration.
- The U.S. Internal Revenue Service is the authoritative source that validates the entity's Taxpayer Identification Number.
- The Defense Logistics Agency is the authoritative source that validates/assigns the Commercial and Government Entity Code.
- The U.S. Small Business Administration validates certification status for 8(a) firms and HUBZones and provides size threshold information.
- Vendor-entered data include core identification data, electronic funds transfer data, points of contact, and assertions, representations, and certifications (if the entity is registering for contracts).
- Ability One validates enterprises that largely employ workers who are blind or have significant disabilities.

SAM is the authoritative source for entity information related to contracting, such as firm revenue that is not publicly available. Entities must validate and update their data annually. See Nancy Y. Moore, Clifford A. Grammich, and Judith D. Mele, *Findings from Existing Data on the Department of Defense Industrial Base*, Santa Monica, Calif.: RAND Corporation, RR-614-OSD, 2014; U.S. General Services Administration, 2014; and U.S. General Services Administration, "System for Award Management," homepage, undated.

reported direct labor hours in its FY 2014 CMRA data, in contrast with the Air Force's 33 percent and the Navy's 6 percent. The other 34 defense agencies reported that 13 percent of ICS dollars were associated with actual contractor-reported direct labor hours in FY 2014 (see Table 3.1).

Because direct labor hours are one of the only unique data fields captured in the ICS, stakeholder views on the necessity and logic of collecting direct labor hours play a considerable role in determining stakeholders' overall sentiments regarding the ICS and CMRA. As with views on the most desirable types of data to enable decisionmaking, views on the utility of direct labor hours were split across functional communities.

Although many individuals with whom we spoke in the manpower and personnel community believed that the collection and reporting of direct labor hours was valuable (as they deemed it to be helpful—or, at least potentially helpful—to inform strategic workforce planning), overall views on the subject were mixed across the community. Five of our 15 manpower and personnel interviewees had positive views of the utility of direct labor hours, whereas three had negative views. Another three did not specify a view, and four were unsure. On the positive side, one former defense official told us that direct labor hours provided a "common unit of measure" to compare the civilian,

Table 3.1
Collection of Actual Contractor-Reported Direct Labor Hours, by DoD Component

Service or Agency	% of ICS Dollars with Contractor-Reported FTEs
DoD-wide	38
Army	100
Navy	6
Air Force	33
Other 34 DoD agencies	15

SOURCE: FY 2014 ICS data.

military, and contractor workforces. Without this, the department was making "dumb, expensive choices."[20] Another interviewee provided a concrete example of how direct labor hours had been used, noting that a defense agency had relied on these data to identify an opportunity for insourcing.[21] Meanwhile, a defense official noted that the ability to compare across the three workforces (military, civilian, and contractor) was useful because it "gives you the full picture." This interviewee added that it was particularly useful for firm fixed-price contracts, into which DoD lacked visibility.[22] Another said that information about "the number of contractors and what they're doing is of great utility."

Two interviewees noted that while direct labor hours were useful, they were not as useful as they could be.[23] One official explained that there were not enough data yet, and the data were not yet integrated into decisionmaking processes.[24] Another OSD official told us that a major problem was that estimated, rather than actual, direct labor hours were "not credible."[25] One personnel official was not sure about the value of direct labor hours, stating that they "may or may not be of benefit." This interviewee worried that without information on types of skills, these data might not be useful.[26] Yet another official in the manpower and personnel community was undecided about the value of direct labor hours, acknowledging the utility of measuring the relative sizes of the civilian, military, and contractor workforces to make total force determinations but claiming to be "still in the mode . . . of formulating an opinion of the usefulness of that data as reported by contractors as opposed to doing algorithmic calculations on our own."[27]

---

[20] Interview with a former defense official, January 2016.

[21] Interview with an OSD official, January 2016.

[22] Interview with a defense official, January 2016.

[23] Interview with an OSD official, January 2016; interview with a defense official, January 2016.

[24] Interview with an OSD official, January 2016.

[25] Interview with an OSD official, January 2016.

[26] Interview with an OSD official, January 2016.

[27] Interview with a defense official, January 2016.

Finally, an official from the manpower world strongly opposed the collection of direct labor hours, calling it a "huge waste of time."[28]

In the acquisition community, there was also a mix of perspectives on the utility of direct labor hours. Two of our interviewees in this community had positive views of the utility of direct labor hours, while five had negative views, one did not specify a view, one was unsure, and three others—who did not speak specifically about direct labor hours—indicated that the CMRA data were not useful yet.

For example, one OSD official voiced the belief that the collection of direct labor hours was valuable, but indirect labor hours would also be useful. This interviewee also noted that much work still needed to be done for direct labor hours to be useful for forecasting.[29] Another acquisition official did not think direct labor hours were accurate enough yet, stating that they were still "a few years away from good data quality." This interviewee reported beginning to use the data in market analysis but said they were of limited value compared with other tools.[30] A third defense official agreed that not enough contractors were reporting direct labor hours to make them statistically valid yet. Nonetheless, this interviewee did see some value in efforts to collect direct labor hours. However, that value was primarily in allowing the department to respond to Congress about contractor numbers and spending. The same interviewee also said of the ICS, "It is used in workforce mix decisions periodically," but pointed out that the system's value is limited by the fact that it is two years late and retrospective.[31]

Two interviewees thought direct labor hours were virtually useless. One noted that in the manpower community, officials are "fixated on the idea that contractors are dirty, moneygrubbing, evil, and we should be insourcing. Whereas the acquisition community isn't as concerned with the number of employees. We're concerned with getting perfor-

[28] Interview with an OSD official, January 2016.

[29] Interview with an OSD official, January 2016.

[30] Interview with a defense official, January 2016.

[31] Interview with a defense official, January 2016.

mance out of industry at fair and reasonable price."[32] Another interviewee said that direct labor hours were a "manufactured requirement we comply with." This interviewee claimed not to "see anyone using [the direct labor hours data] to inform strategic decisions," emphasizing that direct labor hours are data, not information, and the requirement to collect and analyze these data ran counter to instructions to enter into performance-based contracts.[33]

Three of the four individuals in the budgeting community with whom we spoke did not see the value of direct labor hours. One noted that they are "a construct of convenience." Similar to the acquisition officials, this interviewee stated that the budgeting community is "indifferent to the number of people and hours on a contract. It's about accomplishing a defined set of requirements."[34] Another budget official pointed out that because direct labor hours are "just an estimate, you can't directly attribute that to the number of contractors we're buying."[35]

However, one interviewee from this community asserted that collecting direct labor hours from contractors was essential. It is key, the interviewee explained, to compare the components' budget data with the inventory they submit to ensure that budget levels make sense. Without direct labor hours, the contractor FTEs developed by components using different methodologies were not reliable enough for useful comparisons.[36]

All in all, proponents of the reliance on direct labor hours in the ICS as a metric for calculating contractor FTEs focused on their utility for strategic workforce planning and their ease of use for devising contractor FTEs, claiming that the collection of more direct labor hour data will eventually provide a sufficient data pool to enable trend analyses and forecasting to aid in budgetary and strategic workforce

---

[32] Interview with an OSD official, January 2016.

[33] Interview with a defense official, January 2016.

[34] Interview with an OSD official, January 2016.

[35] Interview with an OSD official, January 2016.

[36] Interview with a former OSD official, February 2016.

planning. Opponents focused on the fact that the majority of direct labor hours included in the ICS were estimates based on Army algorithms and therefore only questionably accurate. Opponents further noted that direct labor hours are impossible to audit, even when entered as actuals by contractors themselves, because there is no benchmark against which to compare the data to discern the accuracy of contractor-entered data, other than to look for gross anomalies that stand out in the data set. Opponents complained that ICS data were time-lagged (in that they are not reported for about one year after the FY in question, and they are not updated throughout the year). They also saw the data as retrospective and of little utility for planning or forecasting, primarily because services may be purchased from another contractor at a different price in the future; therefore, even if trend analyses were possible, conducting them would be a futile exercise. Finally, opponents did not find direct labor hours useful for performance-based contracts, in which quality performance of the task, and not the number of workers, is the essential information.

Other Federal Agencies Reported Similar Concerns Regarding the Utility of Direct Labor Hour Data

As noted in Chapter Two, statutory requirements mandate the creation of an analogous service contract inventory for non-DoD federal agencies. This requirement initially mandated in 2008 that OMB establish a pilot program in at least three Cabinet-level departments; the requirement was subsequently extended in 2010 to other civilian agencies.[37] The Consolidated Appropriations Act of 2010, which extended this requirement to other civilian federal agencies, requested elements similar to those in the DoD ICS: description of services purchased, organizational component administering the contract, total dollars obligated and invoiced for the services, contract type and date, contractor name and address, number and location of contractor and subcontractor

[37] Public Law 110-161, 2007.

FTEs for direct labor, and whether the contract was a personal services contract or awarded noncompetitively.[38]

In our interviews with ten individuals who worked for four separate non-DoD federal agencies, we learned more about the process of creating a service contract inventory at non-DoD federal agencies s: Agency employees extract data from FPDS-NG for their required service contract inventory, and contractors enter data, including direct labor hours, into SAM for their supplemental report. This entry of direct labor hours into SAM duplicates efforts by DoD contractors to enter these data into CMRA; in some cases, contractors working for multiple government clients have to enter direct labor hours into both systems for different contracts.

Our intended purpose in speaking with individuals from non-DoD federal agencies was to learn best practices for service contract inventory reporting, but we discovered that because the non-DoD federal service contract inventory requirement was new to these interviewees, they faced many of the same issues and challenges as the DoD officials with whom we spoke. Three of the four non-DoD agencies were only in their second year of reporting for the inventory at the time of our interviews. Meanwhile, two interviewees reported challenges in estimating appropriate costs per direct labor hour, mentioning a problem with the "division" that the system requires. Specifically, they needed to divide cost by the number of FTEs to get the annual cost of FTEs for a contractor employee, but that found that this could provide "horribly skewed results" because "the service encompasses a lot more than labor."[39] We also heard that contractors were unfamiliar with the request to enter data into SAM, and one interviewee reported increasing pushback on the reporting requirement from contractors, particularly larger firms. Another problem we heard was that agencies were unable to generate status reports from SAM themselves, and so they needed to rely on OMB and the U.S. General Services Administration for such reports.

---

[38] Public Law 111-117, 2009.

[39] Interview with a U.S. Department of the Treasury official, January 2016; interview with a U.S. Department of Homeland Security official, March 2016.

Asked how they used the service contract inventory data entered into SAM, interviewees noted that, for the most part, their inventory was a work in progress. As one individual told us, "Right now, we're trying to get to compliance," adding that while the data could be used to look across components in an effort to create efficiencies, the priority for the interviewee's organization was to make sure contractors were reporting the data, and then to make sure the data were valid.[40] Another interviewee reported personally the data for workforce planning, but said that, in general, the data remained an "untapped source of information."[41]

Service Contractor Views on the ICS Reporting Requirement

As noted earlier, we interviewed representatives from nine DoD contracting firms across a variety of sectors and industries. Most of the contractors we interviewed performed work primarily for the federal government, and the majority of the firms also reported that DoD contracts were their primary source of revenue. Of the nine firms—all of which were included in the FY 2014 ICS data on the DPAP website—three did not enter data into the CMRA system, and interviewees from two of those three firms had not even heard of CMRA or the ICS requirement. For these three firms, ICS reporting had not yet been written into their contracts, and their data were estimated in CMRA using algorithmic calculations.

According to interviewees from the six firms that actually reported data to CMRA, we received a variety of responses regarding the amount of time it takes to collect the relevant data and to enter it into the CMRA system. Most respondents said the data entry was fairly easy for them, especially because they had come up with various systems and processes to better automate the data collection and entry. Two interviewees found the CMRA system itself to be very

---

[40] Interview with a U.S. Department of Homeland Security official, March 2016.

[41] Interview with a U.S. Department of the Treasury official, January 2016.

burdensome, however. One interviewee complained that even though the basic information on delivery orders was the same every year, he still has to enter all the data every year: "Once you create a contract, you can have a contract with 100 active delivery orders, so if I work on those 100 delivery orders for five years, I've had to enter that 500 times."[42] We also heard that the CMRA search function did not actually yield results and that the bulk uploader feature, in one instance at least, had been broken for two years.[43]

Contractor interviewees also complained about the multiplicity of reporting requirements. Two noted that they were required to enter a significant amount of overlapping data into CMRA and SAM.[44] Others pointed out that there were four separate "instances" or versions of CMRA (one each for the Army, Air Force, and Navy and one for the other defense agencies), each with its own login information and password, with various inconsistencies across the systems. In addition, several interviewees noted that these were just some of the reporting requirements they faced. As one interviewee explained, "Every additional reporting requirement is a burden for small business."[45]

Other concerns cited in our interviews centered on the proprietary nature of the data reported to CMRA and the difficulty of finding the correct unit identification codes (UICs). In addition, many interviewees simply said that they did not think the data were useful, and they doubted that DoD or Congress used them. As one interviewee said, "I wonder whether anyone is really looking at the data in any meaningful way. I get the sense we're shoveling it into a black hole."[46] Another pointed out that the data being requested was not actually interesting or valuable, and that they obtained the data they needed for their own organizations' planning elsewhere.[47]

---

[42] Interview with a contractor, March 2016.

[43] Interview with a contractor, March 2016.

[44] Interviews with contractors, February and March 2016.

[45] Interview with a contractor, January 2016.

[46] Interview with a contractor, February 2016.

[47] Interview with a contractor, March 2016.

The Future Potential of ICS Data to Meet Congressional and DoD Needs

To summarize, the types of data deemed most relevant for decisionmaking by congressional and DoD stakeholders were (1) those that are analyzed to provide useful information and metrics, rather than just raw data; (2) forward-looking data that can be integrated into budget processes; (3) contractor FTEs to compare with civilian FTEs to inform sourcing decisions; (4) auditable and verifiable data; and (5) data distinguishing between different types of contracts in terms of the values of greatest interest to stakeholders (such as total cost and contractor FTEs). Meanwhile, ICS data are unprocessed, raw data that are time-lagged and retrospective and consist of direct labor hours—and approximately two-thirds of that information is estimated. ICS data are difficult to independently verify, and they report actuals or estimates of direct labor hours for *all* types of service contracts across DoD. A comparison between the ICS data in reality and the data that congressional and DoD stakeholders would find most useful is shown in Table 3.2.

The current gaps between the ICS data and the data most desired for decisionmaking do not necessarily speak to the future potential of

Table 3.2
Current ICS Data Versus Stakeholders' Preferred Data for Decisionmaking

ICS Data	Most Desired Data
Time-lagged and retrospective, though the Army has built a forecasting function into CMRA	Forward-looking and ready to be entered into budget requests during that year's program objective memorandum process
Raw data that mostly have not been analyzed	Analyzed data with trends identified
Direct labor hours (~62% of dollars estimated and ~38% directly entered by contractors)	Contractor FTEs to compare with civilian FTEs to inform insourcing decisions
Difficult to verify independently	Auditable, verifiable data
Reports on numerous types of contracts	Might distinguish among types of contracts in terms of variables of interest (e.g., total cost, contractor FTEs)

the ICS, assuming that plans for the consolidation of the various versions of CMRA into an enterprise-wide CMRA hosted at the Defense Manpower Data Center are allowed to proceed, and assuming that the Total Force Management Support Office in the Office of the Under Secretary of Defense for Personnel and Readiness is established to support this overall effort as planned. Indeed, the Army's longer experience with its own CMRA system indicates that remedies may be available in the longer term for the problem of the data being time-lagged and retrospective, unprocessed, and based primarily on estimates of direct labor hours rather than actual contractor-reported direct labor hours. As mentioned earlier, the Army has developed a process for requiring contractor reporting of actual direct labor hours in CMRA that has allowed 100 percent of Army ICS dollars to be associated with actual contractor-reported direct labor hours, and it has begun to use these data for planning to some degree. As such, Army data are analyzed to a greater extent than ICS data from other defense components, indicating that with time and focus on the development of the CMRA database, several congressional and DoD stakeholder goals for the ICS data may be achievable. The Total Force Management Support Office is one entity that could feasibly take the lead in producing similar trend and forecasting analyses of enterprise-wide CMRA data. Yet concerns about the validity and completeness of the data, variance in the productivity of different contractors and different individual workers, and the failure to distinguish between different types of service contracts in CMRA will likely continue to constrain the potential for such analyses to be robust and able to accurately inform strategic workforce planning or cost/budgeting decisions.

Specifically, two issues are likely to continue hindering the utility of the ICS and CMRA as currently conceived for defense decisionmaking, both of which indicate that the best solution is *not* simply to continue with the development of eCMRA and the process for completing the ICS as planned and required in statute. First, the direct labor hour data will continue to be difficult to independently verify, even when reported directly into CMRA by contractors themselves. There is no way of knowing how each particular contractor measures and tracks direct labor hours; hence, their collection in CMRA is difficult to standardize in a way that helps ensure validity and uniformity across contracts and contractors. Moreover, there are no good data against which to benchmark the (estimated and actual) direct labor hours, other than the direct labor hour data themselves. As discussed earlier, analysts or auditors can search for gross anomalies in the data, but more minor irregularities are likely to go unnoticed.

Second, and more significantly, the requirement to base contractor FTEs on measurements of direct labor hours lacks any language requiring a distinction between different types of contracts for different types of services. The initial congressional intent underlying this requirement appears to have been gaining oversight of the number of deployed operational support contractors in theater and enabling better comparisons of federal civilian personnel and contractors working in staff augmentation roles to identify insourcing opportunities. However, because there is no distinction between different types of contracting, direct labor hours for contractors working in staff augmentation roles are considered to be similar to those of contractors working in complete contracting situations involving the substitution of capital for labor. As discussed in the next chapter, this leads to inaccurate measures and comparisons of labor productivity, resulting in data that are insufficient for budgetary or strategic workforce planning.

CHAPTER FOUR

Why Are There Gaps Between the Current ICS and What Congress and DoD Envisioned?

To gain a comprehensive understanding of the shortcomings of the ICS and why a seemingly simple congressional request has been so difficult to fulfill, it is necessary to step back and consider the economics of defense contracting. Federal sourcing policy starts with a service that the government has determined it needs, and then, within the limits of federal law,[1] it tries to choose the source that can add the most net value in providing that service. The source could be a federal agency, a military organization, a contractor organization, or a nonfederal government agency, among others. Sourcing from nongovernmental organizations is mandated in certain circumstances, as elaborated in Chapter One.[2] In the end, sourcing policy is about choosing an organization to provide goods or services. Each organization considered as a source makes its own decisions about the inputs used to provide the service. As a result, service contractors demonstrate great variability in the degree to which the services they provide replace or augment governmental functions, as well as in the degree to which they substitute capital for labor to optimize their workforce. Consequently, the conversion of a service between a government and nongovernment source varies by contractor and function. This limits the contexts in which direct labor hours can be applied usefully to make efficiency comparisons between different sources. In the next section, we elaborate on

[1] Including, for instance, caps on the total number of military or civilian DoD personnel and prohibitions against contracting out federal government civilian positions.

[2] Executive Office of the President, Bureau of the Budget, 1955; Office of Management and Budget, *Performance of Commercial Activities*, Circular No. A-76, revised May 29, 2003.

these considerations in the context of direct labor hours and their ability to accurately inform sourcing decisions, reflect levels of contracting, and guide workforce planning.

Direct Labor and Sourcing Decisions

Current federal policy specifies the need for comprehensive cost-benefit analyses that include direct labor costs, along with other costs and benefits, to inform significant decisions among alternatives, such as sourcing decisions. OMB Circular A-94 requires a comprehensive accounting of the cost of all inputs and the benefits of service outputs in plausible alternative futures.[3] DoD Instruction 7401.03 translates this guidance for application in DoD.[4] The DoD guidebook on business case analyses of product support requires a similarly comprehensive accounting of the cost of all inputs and benefits of an activity.[5] It addresses plausible futures in a more specific way than Circular A-94, but the approaches are compatible. OMB Circular A-76 calls for a somewhat less comprehensive accounting of costs and benefits in comparing alternative public and private sources, but the approach remains broadly compatible with Circular A-94.[6] Additionally, commercial best practice, while different because of the absence of federal administrative law, also recommends a broad consideration of the costs and benefits in alternative futures.[7]

---

[3] Office of Management and Budget, *Guidelines and Discount Rates for Benefit-Cost Analysis of Federal Programs*, Circular No. A-94, revised October 29, 1992.

[4] U.S. Department of Defense Instruction 7401.03, *Economic Analysis for Decision-Making*, September 9, 2015.

[5] U.S. Department of Defense, *Product Support Business Case Analysis Guidebook*, Washington, D.C., April 2011.

[6] OMB, 2003.

[7] Steven Kelman, *Procurement and Public Management: The Fear of Discretion and the Quality of Government Performance*, Washington, D.C.: AEI Press, 1990; Robin Cooper and Robert S. Kaplan, "Make Cost Right: Make the Right Decisions," *Harvard Business Review*, September–October 1988.

No recommended approach to sourcing singles out the cost of direct labor or counts of direct labor hours for special attention because direct labor typically accounts for only a share of the cost in any source. In FY 2014, the CMRA data show that reported direct labor accounted for an average of 48 percent of total contract costs; this share varied from 27 to 71 percent across one-digit Product or Service Code (PSC) categories, with even greater variance across four-digit PSCs.[8] Similar numbers were not available for government civilian or military organizations, but, if properly measured, direct labor costs typically account for only a fraction of the total costs in those organizations as well.

Consequently, because direct labor does not inform a comprehensive cost-benefit analysis, we found no empirical evidence or concrete examples in our literature search to support specifically using direct labor counts or labor cost as a primary factor in sourcing decisions. Moreover, DoD Instruction 1100.22 specifically prohibits the use of ICS data in making contractor selections. The following sections discuss the limitations of using direct labor hours in sourcing decisions about levels of contracting and workforce productivity and capability planning.

Direct Labor and the Level of Contracting

The level of contracting in which DoD engages varies from *staff augmentation contracting* (also known as *labor contracting*), in which DoD provides the facility, materials, equipment, system, technology, and other inputs to the productive process, to *complete contracting*, in which the contractor provides all productive inputs while DoD provides only contract management.[9] In between these two extremes, there are many opportunities for *mixed contracting*, arrangements that are negotiated and modeled according to circumstances.[10] The most visible type of

---

[8] U.S. Department of Defense, *FY14 CMRA PSC Rates and Factors*, Washington, D.C., 2014.

[9] Sandy Allen and Ashok Chandrashekar, "Outsourcing Services: The Contract Is Just the Beginning," *Business Horizons*, Vol. 43, No. 2, March 2000.

[10] Allen and Chandrashekar, 2000.

contracting is staff augmentation contracting, while complete contracting separates the work entirely from DoD.[11]

The level of management required varies by the level of contracting. Complete contracting requires only high-level program management of total costs and performance by the customer (in this case, DoD), as the contractor provides the employees, materials, processes and systems, technology and equipment, facilities, and supervision.[12] Complete contracting does not involve micromanagement of the inputs employed by the contractor. Contractors are able to make more effective and flexible capital substitutions and staffing decisions than would be the case under an inflexible federal mandate. Mixed contracting and staff augmentation contracting require greater management effort on the part of the customer, as they involve integrating and coordinating multiple workforces within one organization.[13] Some researchers have argued that staff augmentation contracting should be used only to fulfill temporary needs, whereas complete contracting should be used for more long-term needs.[14] Table 4.1 describes these types of contracting and their required levels of management.

Because of the distinction in how staff augmentation contracting, mixed contracting, and complete contracting are managed, the use of direct labor hours for comparison purposes and strategic planning across *all* service contracts in an across-the-board fashion in the ICS is problematic. Even assuming that direct labor hours are valid and precise—a questionable assumption, given the extent to which they are currently estimated and the fact that there is no benchmark against which to check the validity of contractor-reported data—the collection of direct labor hours for complete contracting is inappropriate because each contractor is allowed to determine the productive inputs used to provide the service. Because direct labor hours do not account for dis-

---

[11] Allen and Chandrashekar, 2000.

[12] Allen and Chandrashekar, 2000.

[13] Allen and Chandrashekar, 2000.

[14] CGI Group, *Why Managed Services and Why Not Staff Augmentation? Ensuring Companies Derive the Most Value, Including Flexibility and Skill Access, from IT Service Providers*, Montreal, Quebec, 2015.

Table 4.1
Management Requirements, by Type of Contract

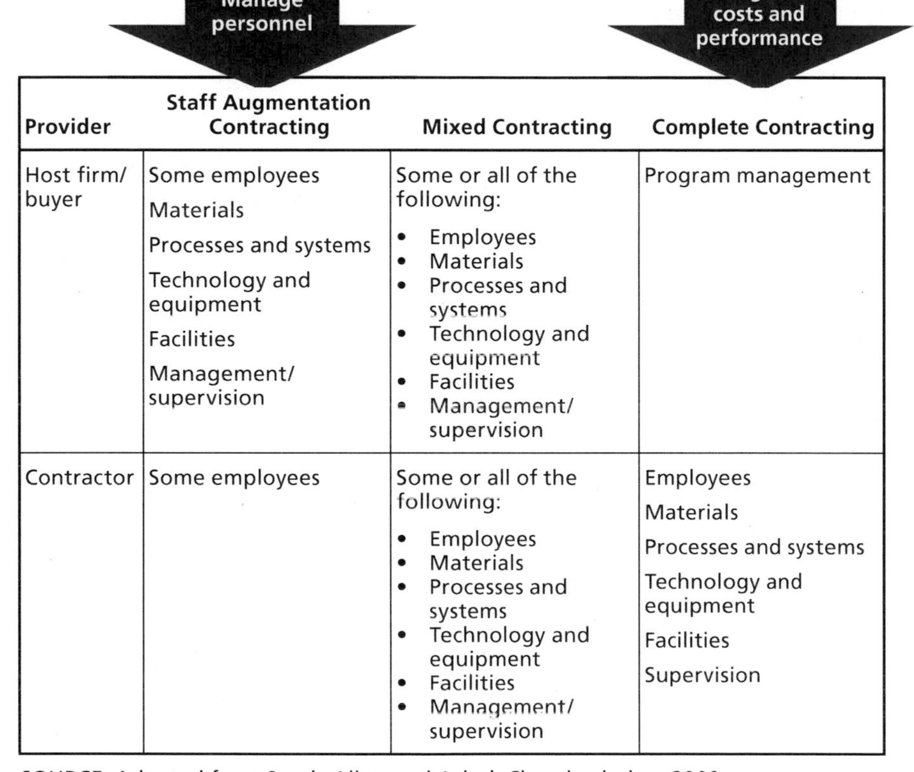

Provider	Staff Augmentation Contracting	Mixed Contracting	Complete Contracting
Host firm/ buyer	Some employees Materials Processes and systems Technology and equipment Facilities Management/ supervision	Some or all of the following: • Employees • Materials • Processes and systems • Technology and equipment • Facilities • Management/ supervision	Program management
Contractor	Some employees	Some or all of the following: • Employees • Materials • Processes and systems • Technology and equipment • Facilities • Management/ supervision	Employees Materials Processes and systems Technology and equipment Facilities Supervision

SOURCE: Adapted from Sandy Allen and Ashok Chandrashekar, 2000.

tinctions between the various types of contracting, they are insufficient for quality strategic workforce planning.

Proponents of the ICS approach of collecting direct labor hours for all contracts might argue that this does not matter, as one of the ultimate goals of the ICS is to inform insourcing decisions. However, comparing per-person costs as required in insourcing decisions is necessary only for staff augmentation contracting, in which DoD provides the facility, materials, equipment, system, technology, and other inputs to the production process. Other contracting arrangements may be enhanced by capital or technology investments that increase the efficiency of their labor, but that also increase per-person costs, a concept that we explore in the next section.

Direct Labor and Workforce Planning

The potential benefits that may be incurred by contracting non–inherently governmental functions to private-sector providers begin with either reducing the cost or increasing the value of the service contracted. This is achieved by increasing the focus on core competencies and enhancing the quality of service provision.[15] While not always the case because of imperfect competition among competing contractors and inaccurate cost budgeting,[16] contracting services to private-sector organizations may be beneficial in terms of productivity and capability enhancement, as well as lower costs. Research demonstrates that because the government's personnel and compensation systems are slow and rigid, advancements can be divorced from achievement, and attracting and retaining talented personnel is often difficult.[17] Consequently, the productivity and capability of government personnel may be lower than in the private sector. Additionally, research demonstrates that, because of the greater deferred costs, higher noncash benefits, and the costs of required support staff, total compensation for federal civilians may be up to 40 percent higher than that of equally skilled

---

[15] Robert Klitgaard and Paul C. Light, eds., *High-Performance Government: Structure, Leadership, Incentives*, Santa Monica, Calif.: RAND Corporation, MG-256-PRGS, 2005.

[16] U.S. Small Business Administration, *Government Contracting 101: A Guide for Small Businesses, Part 1—Small Business Contracting Programs, Supplemental Workbook*, January 2012; Albert A. Robbert, Susan M. Gates, and Marc N. Elliot, *Outsourcing of DoD Commercial Activities: Impacts on Civil Service Employees*, Santa Monica, Calif.: RAND Corporation, MR-866-OSD, 1997.

[17] David M. Walker, Comptroller General of the United States, *Results-Oriented Government: Shaping the Government to Meet 21st Century Challenges*, statement before the Subcommittee on Civil Service and Agency Organization, Committee on Government Reform, U.S. House of Representatives, Washington, D.C.: U.S. General Accounting Office, September 2003; Kay Coles James, *A Fresh Start for Federal Pay: The Case for Modernization*, Washington, D.C.: Office of Personnel Management, April 2002; Beth J. Asch, *The Pay, Promotion, and Retention of High-Quality Civil Service Workers in the Department of Defense*, Santa Monica, Calif.: RAND Corporation, MR-1193-OSD, 2001.

private-sector employees in identical jobs.[18] This gap would seem to imply that nongovernment personnel can provide the same services at lower costs; however, the research is far from clear-cut on this issue. Rather, studies comparing federal and private-sector compensation have arrived at vastly different conclusions, usually depending on the methodology used.[19] The most rigorous studies tend to show a pay premium for federal workers with lower levels of educational attainment and a pay penalty for federal workers with higher levels of educational attainment.[20]

Congress has recommended using the ICS to inform workforce planning regarding productivity, capacity, and scale.[21] However, without precise measures of productivity and capability—the determinants of which can vary across individuals, organizations, and sectors—the ICS is ill equipped to address workforce planning. Data on direct labor in the ICS describe capability and productivity only for the staff augmentation contracts because counting the number of people working in an organization is not the same as measuring the net value generated per person or the capabilities of each person.[22]

---

[18] James Sherk, *Inflated Federal Pay: How Americans Are Overtaxed to Overpay the Civil Service*, Washington, D.C.: Heritage Foundation, CDA10-05, July 2010; Chris Edwards, "Reducing the Costs of Federal Worker Pay and Benefits," *Downsizing the Federal Government*, September 20, 2016; George J. Borjas, *The Wage Structure and the Sorting of Workers Into the Public Sector*, Cambridge, Mass.: National Bureau of Economic Research, Working Paper 9313, October 2002; Justin Falk, *Comparing the Compensation of Federal and Private Sector Employees*, Washington, D.C.: Congressional Budget Office, January 2012.

[19] David H. Bradley, *Comparing Compensation for Federal and Private Sector Workers: An Overview*, Washington, D.C.: Congressional Research Service, July 30, 2012.

[20] Bradley, 2012; Falk, 2012; U.S. Department of Labor, Executive Office of the President, and Office of Personnel Management, *Report on Locality-Based Comparability Payments for the General Schedule: Annual Report of the President's Pay Agent*, Washington, D.C., 2014; Andrew Biggs and Jason Richwine, *Comparing Federal and Private Sector Compensation*, Washington, D.C.: American Enterprise Institute, Economic Policy Working Paper 2011-02, March 4, 2011; Edwards, 2016.

[21] U.S. House of Representatives, Committee on Armed Services, 2010, p. 271.

[22] Note, however, that the collection of direct labor hour data pertaining to operational contract support may be necessary to increase congressional oversight of the number of deployed contractors working on DoD contracts in theater. This subset of all DoD contracts can and

While productivity is easy to define in principle (the ratio of output to input), in reality, it is determined by a variety of inputs and can be difficult to measure at disaggregated levels.[23] Economic theory indicates that physical capital (e.g., machinery, computers) and human capital (e.g., education, training, experience, governance structure) are among the inputs for labor productivity.[24] Consequently, direct labor substitutions between different components of the total force—military, civilian, and contractor—are not always one for one within or across sectors because of individual-, organization-, and sector-level variation and gaps in productivity inputs.[25] Gaps in productivity are reflected in costs and also complicate the comprehensive accounting of costs and benefits required to inform the sourcing decision. Investments or depreciations that change the level of capital also alter the productivity of labor in processes that require their interaction. Firm- and sector-level differences in nonlabor inputs produce gaps in wages that may not reflect what the same individual may have produced in a different firm or sector.[26] Additionally, complementarities in produc-

---

most likely should be treated differently from other service contracts, as elaborated in Chapter Seven.

[23] Erik Brynjolfsson and Adam Sanders, *Wired for Innovation: How Information Technology Is Reshaping the Economy*, Cambridge, Mass.: MIT Press, 2013.

[24] Daron Acemoglu and David Autor, *Lectures in Labor Economics*, Cambridge, Mass.: Massachusetts Institute of Technology, 2016.

[25] Moreover, prior research on commercial best practices has found no evidence that firms in the commercial sector collect direct labor hours, indicating that this practice is not worthwhile in this sector. See, for instance, Robert M. Monczka, Robert J. Trent, and Robert B. Handfield, *Purchasing and Supply Chain Management*, 2nd ed., Cincinnati, Ohio: South-Western College Pub., 2002; Michiel R. Leenders, Harold E. Fearon, Anna E. Flynn, and P. Fraser Johnson, *Purchasing and Supply Management*, 12th ed., New York: McGraw-Hill Irwin, 2002; Donald W. Dobler and David N. Burt, *Purchasing and Supply Management: Text and Cases*, 6th ed., New York: McGraw-Hill, 1996; Nancy Y. Moore, Clifford A. Grammich, and Robert Bickel, *Developing Tailored Supply Strategies*, Santa Monica, Calif.: RAND Corporation, MG-572-AF, 2007; Nancy Y. Moore, Laura H. Baldwin, Frank A. Camm, and Cynthia R. Cook, *Implementing Best Purchasing and Supply Management Practices: Lessons from Innovative Commercial Firms*, Santa Monica, Calif.: RAND Corporation, DB-334-AF, 2002.

[26] Kenneth J. Arrow, H. B. Chenery, B. S. Minhas, and R. M. Solow, "Capital-Labor Substitution and Economic Efficiency," *Review of Economics and Statistics*, Vol. 43, No. 3, August 1961.

tive inputs—including human capital inputs, such as education and training—mean that *the same individual may not be equally productive in different firms or sectors.*[27] Relatedly, contractor staffing changes may reduce the number of direct labor hours but achieve the same productivity. The number and mix of contract workers and their skills may differ from the allocation of federal government staff. Higher-quality organizations can produce a given level of service at a lower cost because of higher labor productivity, justifying higher per-person costs. In short, while productivity is important, it is very difficult to measure, and simply measuring the number of workers or the number of hours worked does not inform workforce productivity.

Productivity is also related to workforce planning in terms of the number of workers required to perform a given function. If productivity were maximized, it would imply that an organization has employed the "right" number of workers. However, because productivity is difficult to measure, and the available measures (direct labor hours) do not inform productivity, maximum productivity cannot be determined. While there are potentially other measures—like work backlogs—that could be used to assess whether the optimal number of workers are employed for specific functions, these measures are not collected in the ICS. Consequently, the ICS does not effectively inform workforce planning.

In addition to the number of workers, workforce planning also involves capability planning to ensure the future availability of skills for contracted services and for inherently governmental services (those for which there is no adequate private-sector capability).[28] For example, military organizations embed contractor employees in military units to provide skills that the military organizations have chosen not to create or sustain in the military workforce. These employees are often veterans who mentor junior technical support military personnel. They may also support recently fielded military hardware or provide the skills required to ensure computer support and connectivity to tactical

[27] Victor A. Beker and Esteban Albisu, "Raw Labor: Homogeneous or Heterogeneous?" April 29, 2010.

[28] Office of the Under Secretary of Defense for Acquisition and Technology, 1996, p. 53.

units. For similar reasons, civilian organizations embed both military and contractor employees in roles that are typically tied more to office-based tasks. However, the ICS does not contain the relevant information to inform capability plans. Variations and gaps in productivity reflect variation in workforce capability. Because historical and current contracting decisions do not fully inform future needs, and because the ICS conflates labor and nonlabor costs and does not precisely identify the sources of capabilities, the ICS cannot inform strategic workforce planning efforts despite containing information on the provision of services that have been or are currently contracted.

CHAPTER FIVE

Potential Insights from Other Data Sources

As discussed in Chapter One, one of the objectives of this study was to develop alternative methods for collecting, processing, and reporting data on contracted services to meet the ultimate goals of the ICS and to facilitate DoD's strategic planning and decisionmaking efforts regarding manpower, budgeting, and the acquisition of services. Specifically, we explore how existing databases or IT systems present a timely solution and provide data relevant to strategic workforce planning. In this chapter, we explore insights relevant to policymakers' goals for the ICS from FPDS-NG and budget data.

The FPDS-NG provides, for contract actions of at least $3,000, information on the amount of the contract action, identification codes indicating whether the firm providing the service is a small business, the North American Industry Classification System (NAICS) code for the service being purchased, and the PSC, a more finely grained indicator than NAICS codes of the nature of the goods and services purchased.[1] In this chapter, we summarize insights from purchases of services using the PSC for the contract action. While new contract data in the public version of the FPDS-NG is subject to a 90-day delay for security and verification purposes, the government version can provide policymakers with a relatively current overview of DoD's purchases of services. Specifically, FPDS-NG data can help identify the characteris-

[1] For more on PSCs, including a list, see U.S. General Services Administration, *Federal Procurement Data System Product and Services Code Manual*, Washington, D.C., August 2015. Using the PSC, we can identify contract actions for (or purchases of) services that are of interest to Congress.

tics of services that DoD purchases, the growth in these services over time and which types of services have seen the greatest growth, and the means that DoD uses to procure services (e.g., firm fixed-price contracts, performance-based contracts). These data, when coupled with budget projections, can illustrate likely changes in total service spending in the coming years.

Characteristics of Service Categories

FPDS-NG data offer a broad overview of the types of services that DoD purchases. Table 5.1 shows the distribution of dollars spent by DoD on services, by PSC category, in FY 2015.[2] There are nearly 5,000 PSCs in total, but focusing first on these 24 categories provides a broad indication of how DoD allocates its dollars for services. For each PSC category, we show the category description, the percentage of all contracted service dollars in that category, the percentage of dollars in the category that go to small businesses, the number of agencies and contracting offices, and the number of contracts used to purchase the services.

As shown, half of all DoD service purchases are in the top three PSC categories by dollars, three-fourths are in the top six categories, and nine out of ten are in the top nine. Nearly one-fourth of DoD spending on services goes to the broad category of Support (Professional/Administrative/Management). This may indeed include the use of service contracts to augment DoD staff, which has been of concern to some in Congress. The need for such services is widespread, with nearly two dozen agencies and nearly 700 contracting offices purchasing them through more than 18,000 contracts.

Yet, while only a contracting officer with access to both the contract specifics and a clear definition of *staff augmentation services* would

---

[2] PSC categories are groupings of similar PSCs. For example, PSC category A, Research and Development, includes such services as agriculture, defense, housing, social services, transportation, and research on still other topics. Among these PSCs, as we address in discussing specific PSCs, are those for aircraft operational systems and other advanced development of defense systems.

Table 5.1
DoD PSC Category Spending Characteristics, FY 2015

PSC Category	Description	% of All Dollars for Contracted Services	% of Dollars for Category of Services Going to Small Business	Number of Agencies	Number of Contracting Offices	Number of Contracts
R	Support (Professional/Administrative/ Management)	23	29	23	688	18,180
A	Research and Development	16	18	18	357	14,040
J	Maintenance, Repair, and Rebuilding of Equipment	12	13	22	637	17,395
Q	Medical	9	10	14	269	3,191
D	Information Technology and Telecommunication	8	34	23	538	14,189
Y	Construction of Structures/Facilities	7	35	12	286	5,481
Z	Maintenance, Repair, Alteration of Structures/ Facilities	6	57	17	379	9,238
V	Transportation/Travel/Relocation	4	18	18	450	4,739
S	Utilities and Housekeeping	4	29	20	473	10,960
U	Education/Training	2	23	21	441	3,987
M	Operation of Structures/Facilities	2	17	12	203	932
F	Natural Resources Management	2	41	13	295	2,339

Table 5.1—Continued

PSC Category	Description	% of All Dollars for Contracted Services	% of Dollars for Category of Services Going to Small Business	Number of Agencies	Number of Contracting Offices	Number of Contracts
C	Architect and Engineering Services	2	30	11	256	2,295
B	Special Studies/Analysis, not R&D	1	21	20	294	1,158
K	Modification of Equipment	1	7	8	180	415
L	Technical Representative	1	9	12	232	779
N	Installation of Equipment	<0.5	33	18	358	1,349
H	Quality Control, Testing, and Inspection	<0.5	22	17	368	1,610
W	Lease/Rental of Equipment	<0.5	37	18	397	4,122
X	Lease/Rental of Structures/Facilities	<0.5	10	16	235	1,004
T	Photo/Map/Print/Publication	<0.5	51	15	176	389
P	Salvage	<0.5	52	8	104	244
G	Social	<0.5	92	13	221	1,373
E	Purchase of Structures/Facilities	<0.05	69	7	31	51

SOURCE: FY 2015 FPDS-NG awards (data as of March 2016).

NOTE: Categories listed by rank as a share of overall spending on contracted services.

be able to tell for certain, many of these services are very likely not staff augmentation. Note, for example, that 7 percent of spending is for the construction of structures and facilities, services that the private sector can provide. In addition, 4 percent is for utilities and housekeeping, with utilities being another necessary service that the government is not likely to provide itself. How well the government can leverage these purchases is not immediately clear, given that many firms providing construction or utility services are likely local businesses.

Small-business purchases may be of concern to policymakers should they wish to uphold small-business purchasing goals—which were 21.35 percent for FY 2014, 21.60 percent for FY 2015, and 21.26 percent for FY 2016—while seeking other ways to leverage the supply base.[3] DoD uses thousands of contracts to purchase services in many of these categories, indicating that there may be opportunities to leverage purchases through contract consolidation. However, the high proportion of small businesses in some of these categories suggests that any consolidation may require caution. For example, DoD uses more than 9,000 contracts to purchase services in the Maintenance, Repair, Alteration of Structures/Facilities category, and it spends 57 percent of these dollars with small businesses. This is nearly three times the department goal for small-business purchases, indicating that DoD may be using this category to boost its overall small-business spending. Any leveraging of spending through contract consolidation should examine specific services in this category, particularly whether they are best provided by small or even local firms.

---

[3] On small-business purchasing goals, see Department of Defense Office of Small Business Programs, "Small Business Program Goals," web page, undated. For more on the trade-offs policymakers may encounter between best purchasing practices and small-business goals, as well as the varying penetration of small businesses in different industries and categories of DoD purchases, see Nancy Y. Moore, Clifford A. Grammich, Julie DaVanzo, Bruce J. Held, John Coombs, and Judith D. Mele, *Enhancing Small-Business Opportunities in the DoD*, Santa Monica, Calif.: RAND Corporation, TR-601-1-OSD, 2008.

Growth by Service Categories

FPDS-NG data can also provide insights on where growth is occurring. Figure 5.1 shows DoD's 2014 spending in leading PSC categories in 1999, 2008, and 2015 in constant dollars over time. We chose 1999 as our first point of comparison because it predates both U.S. military operations in the wake of the September 2001 terrorist attacks and changes in reporting practices that would raise comparability problems in some years. We chose 2008 as our second point of comparison because it was the year when DoD spending peaked in constant dollars.[4] We chose 2015 as our third point of comparison because it was the most recent year for which complete FPDS-NG data were available at the time of our research.

From 1999 to 2008, when DoD's overall spending grew by 87 percent in real terms, most service categories saw even faster growth.[5] For example, spending in the category Construction of Structures/Facilities grew nearly fivefold, while that for Installation of Equipment grew more than fourfold and that for Support (Professional/Administrative/Management) more than tripled. From 2008 to 2015, overall DoD spending decreased by 25 percent, and spending in nearly all categories of services also decreased. Nevertheless, in some PSC categories, spending did not decrease as quickly as overall DoD spending—and, in some cases, it even increased. For example, spending in the Medical category has increased in recent years, and Transportation/Travel/Relocation spending was virtually unchanged during this period. Thus, it is unlikely that contracts in these categories were for staff augmentation. Overall, from 1999 to 2015, DoD expenditures increased 41 percent in real terms, a rate exceeded by eight categories of services. Four categories of services grew during this period, but not as quickly, and 12 categories of services saw decreases in real spending. Altogether, spend analyses of FPDS-NG data can show which categories of services likely increased because of extraordinary

---

[4] Office of the Under Secretary of Defense, Comptroller, 2016.

[5] Office of the Under Secretary of Defense, Comptroller, 2016.

Figure 5.1
DoD Spending in Leading PSC Categories, FYs 1999, 2008, and 2015

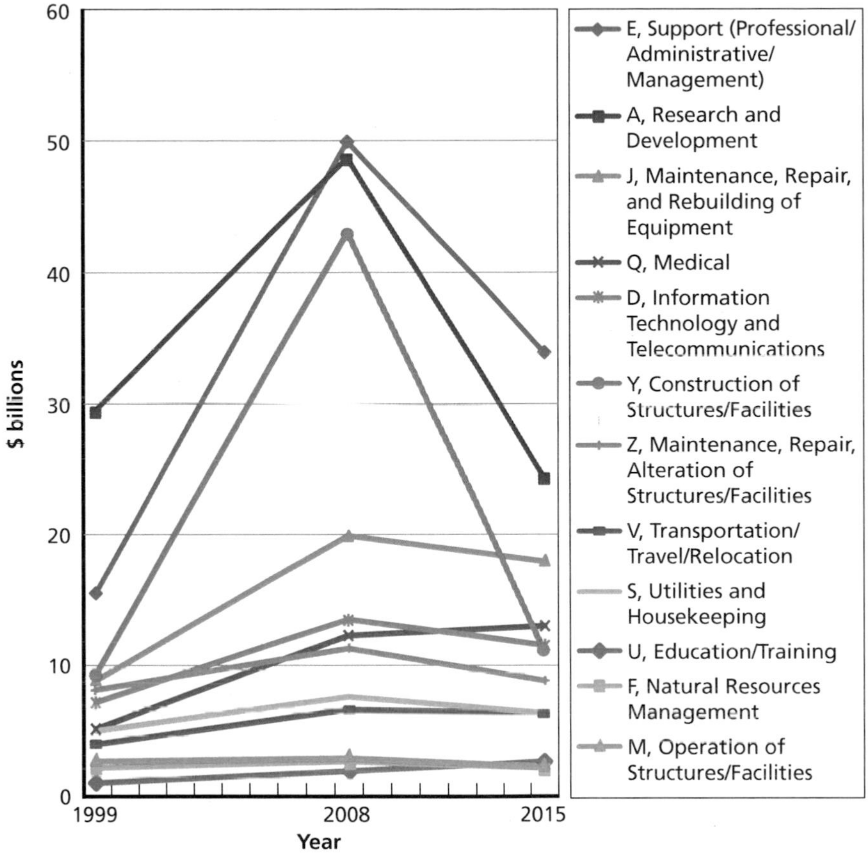

SOURCE: FPDS-NG awards (data as of March 2016).
NOTE: Data are in constant FY 2017 dollars. Because the threshold for reporting contract actions to FPDS in FY 1999 was $25,000, we limited our analysis to contracts with actions of at least $25,000. This had little effect on our results. In FY 2015, contracts with actions of at least $25,000 accounted for more than 99.5 percent of DoD expenditures in these service categories.
RAND *RR1704-5.1*

events, such as military action in Iraq and Afghanistan, and which categories are a source of concern when spending continues to rise.

Characteristics of Specific Services

We gleaned further insights from the FPDS-NG data by examining spending within specific PSCs. For each of the ten PSCs on which DoD spent the most money in FY 2015, Table 5.2 lists the value, the number of contracts and contractors, the number of agencies purchasing services, the number of industries from which DoD purchased the services, and the percentage of expenditures going to small businesses.

It is not surprising that Support (Professional/Administrative/Management), the PSC category on which DoD spent the most money in FY 2015, accounted for four of the top seven PSCs that year. Yet, the characteristics of each of these individual PSCs differ markedly. Many purchases of Support—Professional: Engineering/Technical (R425) may have been for specialized capabilities not available within DoD rather than for staff augmentation. Such purchases were also quite widespread and diverse, with 1,908 contractors providing services to 16 DoD agencies through 3,591 contracts in 154 industries. Support—Professional: Other (R499) spending was even more diverse, with 2,914 contractors in 277 industries providing services to 22 agencies through 4,743 contracts. Many of these services may include some form of staff augmentation, but identifying which services qualify as staff augmentation would require further study. Support—Management: Logistics Support (R706) and Support—Professional: Program Management/Support (R408) are also diverse PSCs, each covering services in about 100 industries in FY 2015. The greater number of contracts than contractors for these services, as well as the large number of agencies purchasing these services, points to potential opportunities to leverage spending, though the proportion of spending going to small businesses suggests that such leveraging must be approached with care.

Results from some of these categories of services suggest that DoD is not likely augmenting staff—and it is possibly already lever-

Table 5.2
DoD Spending on Services, FY 2015, by PSC

PSC	Description	$ Millions	Contracts	Contractors	Agencies	Industries	% Small Business
R425	Support—Professional: Engineering/Technical	12,696	3,591	1,908	16	154	29
Q201	Medical—General Health Care	10,448	508	271	10	39	4
R499	Support—Professional: Other	6,427	4,743	2,914	22	277	29
R706	Support—Management: Logistics Support	4,467	1,408	618	17	113	10
J015	Maintenance, Repair, and Rebuilding of Equipment—Aircraft and Airframe Structural Components	4,395	335	164	4	24	5
D399	IT and Telecom—Other IT and Telecommunications	4,287	2,426	1,297	22	102	38
R408	Support—Professional: Program Management/Support	3,096	1,815	1,137	19	95	42
J016	Maintenance, Repair, and Rebuilding of Equipment—Aircraft Components and Accessories	2,317	428	250	6	48	19
S216	Housekeeping—Facilities Operations Support	2,275	1,097	784	14	119	31
Y1JZ	Construction of Miscellaneous Buildings	2,101	1,068	776	4	39	36

SOURCE: FY 2015 FPDS-NG awards (data as of March 2016).

aging spending to a considerable extent. Spending in FY 2015 on IT and Telecom—Other IT and Telecommunications (D399) was spread across a large number of contracts and contractors, and 102 industries; much of this spending was with small businesses as well. It also appears that spending on Medical—General Health Care (Q201), Maintenance, Repair, and Rebuilding of Equipment—Aircraft and Airframe Structural Components (J015), and Maintenance, Repair, and Rebuilding of Equipment—Aircraft Components and Accessories (J016) was relatively consolidated and perhaps already leveraged.

Growth in Specific Services

Data on changes in spending at the PSC level over time offer further insights. However, because of the changes in PSC definitions, this analysis is somewhat limited. Figure 5.2 shows, for the ten PSCs on which DoD spent the most money in FY 2015, the amount it spent in constant dollars in 1999, 2008, and 2015.

Spending in the top nine PSCs in which DoD spent the most in FY 2015 grew from 1999 to 2008. Approximately half of the top PSCs decreased in spending from 2008 to 2015, just as overall DoD spending decreased. Nevertheless, other PSCs increased in spending during this time frame. For example, spending in Maintenance, Repair, and Rebuilding of Equipment—Aircraft Components and Accessories (J016) increased more than two-thirds from 2008 to 2015. Areas of growth may warrant further investigation to identify the reasons for growth in DoD service spending, though it is critical for analysts to understand the changes in PSCs and their definitions over time. Changes in spending for the tenth PSC, Construction of Miscellaneous Buildings (Y1JZ), were not available because that PSC was created after 2008.

Figure 5.2
DoD Spending in Top Ten PSCs, FYs 1999, 2008, and 2015

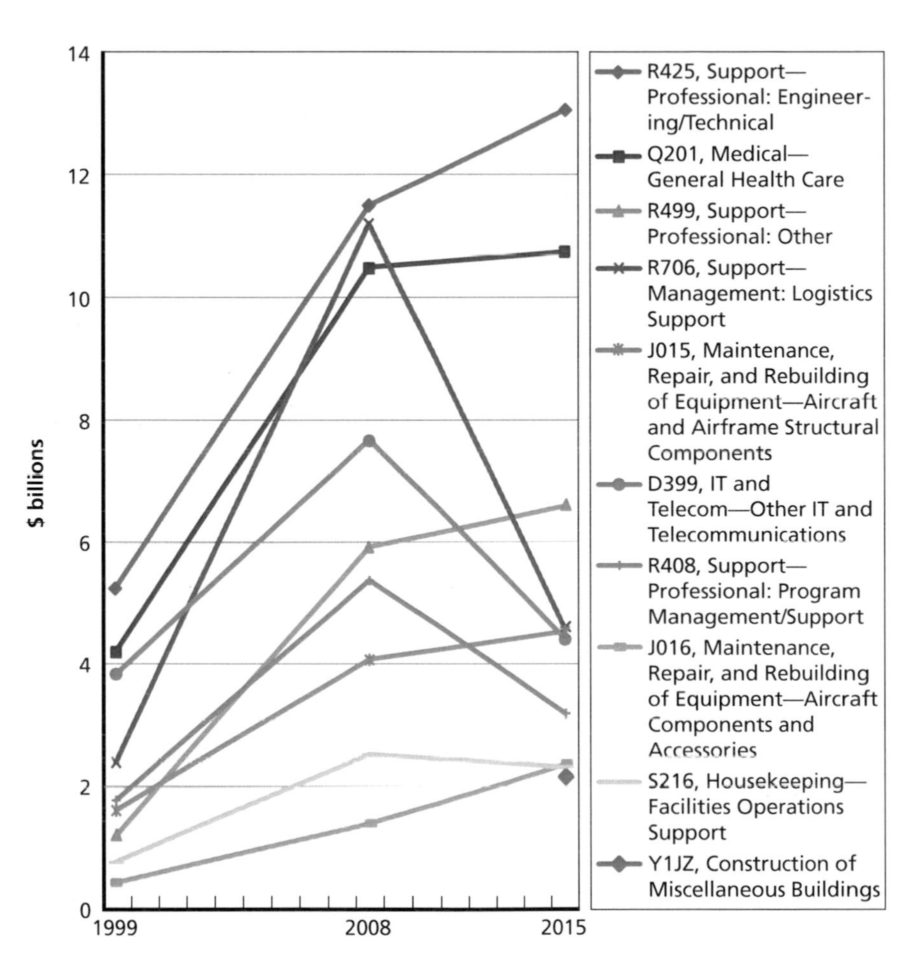

SOURCE: FPDS-NG awards (data as of March 2016).
NOTE: Data are in constant FY 2017 dollars. Because the threshold for reporting contract actions to FPDS in FY 1999 was $25,000, we limited our analysis to contracts with actions of at least $25,000.
RAND *RR1704-5.2*

Service Contract Practices

FPDS-NG data can also offer insights into how DoD executes its service contracts. Figure 5.3 shows the proportion of all contracts and dollars in the top eight DoD service contract types in FY 2015 by dollar amount obligated, as well as the proportion of small-business contracts by contract type. Nearly 90 percent of contracts, and just over half of contract dollars (and more than 70 percent of small-business dollars) went to firm, fixed-price contracts. While DoD has encouraged the use of such contracts, these contracts have fewer contract line-item num-

Figure 5.3
DoD Service Contracts, by Contract Type

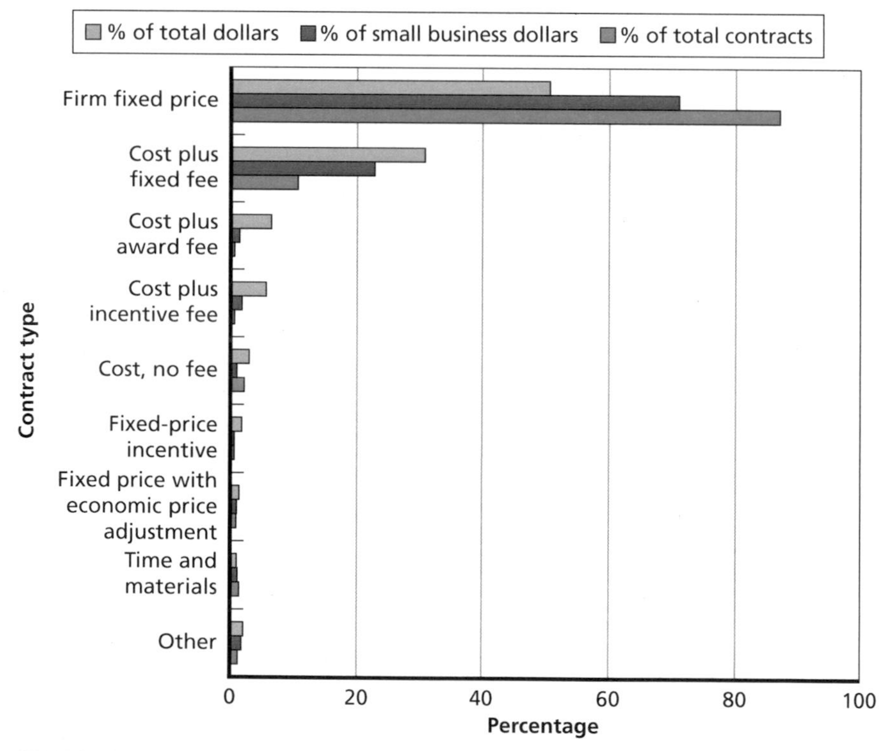

SOURCE: FY 2015 FPDS-NG awards (data as of March 2016).
RAND *RR1704-5.3*

bers, and, hence, allow less precision in identifying how contractors spend the money they receive for services.[6] Such contracts also do not include performance incentives, as others may.

FPDS-NG data also offer insights into the extent of competition for DoD services. Figure 5.4 shows the extent to which actions coded with a service PSC on DoD contracts in FY 2015 were open to competition. Only 29 percent of DoD service contracts were subject to

Figure 5.4
DoD Service Spending, by Contract Type

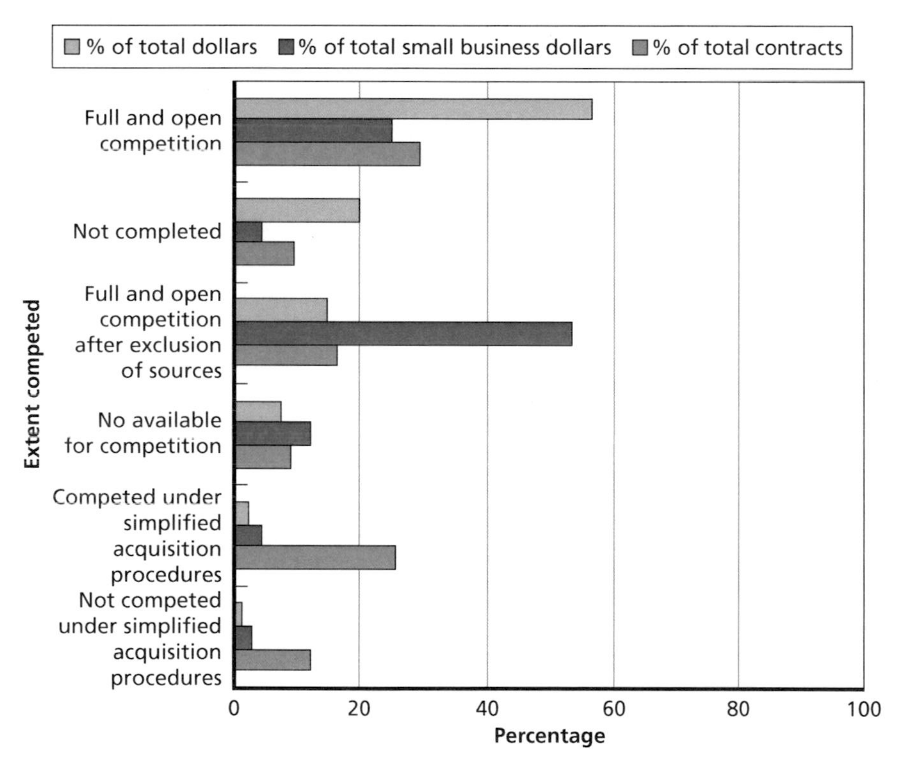

SOURCE: FY 2015 FPDS-NG awards (data as of March 2016).
RAND *RR1704-5.4*

---

6 See, for example, U.S. Department of Defense, Office of the Inspector General, *Contracts for Professional, Administrative, and Management Support Services*, Washington, D.C., Report No. D-2004-015, October 30, 2003, and U.S. Department of Defense, *Time-and-Materials and Labor-Hour Contracts: The New Policies*, Washington, D.C., 2006a.

full and open competition, and 16 percent were subject to full and open competition only after exclusions, typically those to favor small businesses.[7] Nevertheless, when considering how DoD contract dollars were spent on services, we found that 56 percent went to contracts that were procured through full and open competition.

Projecting Future Growth in Service Spending

Finally, FPDS-NG data, when coupled with budget projections, can offer some insight into the likely magnitude of future spending for services. As Table 5.3 shows, in FY 2015, about three-fourths of DoD's contracts for services, and about two-thirds of spending on services, were in the operations and maintenance (O&M) budget category—proportions that have not changed substantially in recent years.

Congress appropriates DoD's budget by category, and constrains moves between categories. As a result, using categories of budget projections can help indicate where DoD budget reductions are likely to occur—and how they will affect spending on services. Figure 5.5 indicates that O&M has been the focus of DoD budget reductions in recent years—and that these reductions will likely continue in FY 2017 before leveling off.

Applying the percentage distribution of spending for services to each budget category as listed in the National Defense Budget Estimates yields an estimate of the likely total amount that DoD will spend on services in coming years.[8] As shown in Table 5.4, the "% of Budget Spending on Purchases" column indicates the percentage of spending in each category that went to purchases (rather than personnel), including services, relative to the amount budgeted for that purpose in FY 2015. It indicates, for example, that 65 percent of DoD's O&M budget went to purchases, and that, across DoD, $282 billion

---

[7] As Figure 5.4 shows, 53 percent of all DoD dollars spent with small businesses for services were through contracts that had full and open competition after exclusion of sources.

[8] Office of the Under Secretary of Defense, Comptroller, 2015.

Table 5.3
DoD Service Spending, by Budget Category, FYs 2012–2015

Budget Year	**2012**		**2013**		**2014**		**2015**	
Total service spending ($ billions)	186		161		156		144	
	%		%		%		%	
Budget Category	**Contracts**	**$**	**Contracts**	**$**	**Contracts**	**$**	**Contracts**	**$**
O&M	74	64	74	65	75	65	72	66
RDT&E	13	17	13	17	13	17	13	17
Procurement	2	6	2	7	2	7	2	6
Military construction	4	8	4	6	4	6	3	5
Non-DoD, U.S. Army Corps of Engineers	7	3	7	3	7	3	7	3
Non-DoD, other	2	1	1	2	1	1	1	2
Military personnel	1	1	1	1	1	1	1	1
Family housing	1	<1	1	<1	1	<1	<1	<1
Total	100	100	100	100	100	100	100	100

SOURCE: FPDS-NG awards, FYs 2012–2015 (data as of March 2016).

Figure 5.5
Largest Budget Reductions in O&M

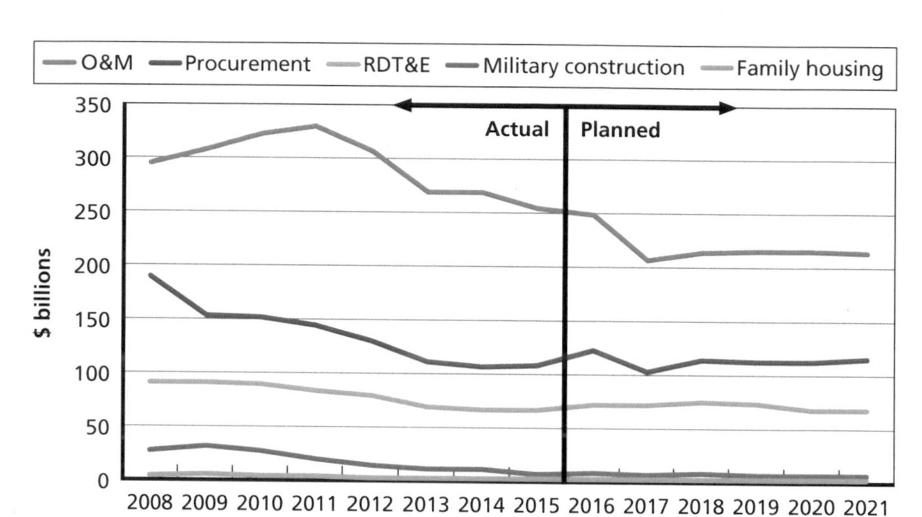

SOURCE: Office of the Under Secretary of Defense, Comptroller, 2016, Table 6-1.
NOTE: Data are in constant FY 2017 dollars.
RAND *RR1704-5.5*

in FY 2015 was for purchases.[9] The column "% of Budget Spending on Purchases Allocated to Services" indicates the percentage of purchases that were for services as identified in the FPDS-NG. It indicates, for example, that 59 percent of O&M purchases were for services, and that, in total, DoD spent $148 billion on services in FY 2015. This was a decrease of $38 billion from the $186 billion spent on services in FY 2012, as shown in the row "Decrease from 2012."

Assuming these rates are constant and applying them across projected spending by budget categories in future years, we were able to estimate future DoD expenditures for services. The results indicate that DoD spending on services will decrease to $123 billion by FY 2017.

[9] Because construction projects may require years to arrange all permits and permissions, expenditures for them may occur several years after they are authorized. We surmise that this is why military construction service expenditures were 115 percent of what was budgeted in FY 2015, as shown on the figure.

Table 5.4
Decreases in Service Spending and O&M Spending

Budget or Appropriation Category	% of Budget Spending on Purchases	% of Budget Spending on Purchases Allocated to Services	Estimated Spending on Services ($ billions)					
	2015	2015	2016	2017	2018	2019	2020	2021
O&M	65	59						
Procurement	52	16						
RDT&E	52	73						
Military construction	115	99						
Military personnel	1	98						
Family housing	33	94						
Total	$282 billion	$148 billion	$143 billion	$123 billion	$130 billion	$129 billion	$126 billion	$126 billion
Decrease from 2012 ($186 billion)	NA	$38 billion	$43 billion	$63 billion	$56 billion	$57 billion	$60 billion	$60 billion
Service decrease goal				$4.1 billion				

SOURCES: FY 2015 FPDS DoD data (as of March 2016); Office of the Under Secretary of Defense, Comptroller, 2016, Table 6-1.
NOTE: Table uses FY 2016 constant spend/service rates.

While it will increase modestly after that, we estimate that DoD will see reductions of $60 billion from what it spent on services in FY 2012.

Such savings would help DoD far exceed congressional goals for FYs 2012–2017 that it achieve savings in its civilian and contractor workforces at least as large in percentage terms as those in military basic pay resulting from reductions in military end strength. A GAO study suggested that this would require a $4.1 billion reduction in service spending, shown in the 2017 column in the row "Service Decrease Goal."[10] As noted, our analyses suggest that, by FY 2015, DoD had already realized $38 billion in savings in service spending since FY 2012, far exceeding the $4.1 billion by which Congress mandated it reduce its spending on contracted services.

Using FPDS-NG Data to Generate Insights on Contracted Services

As illustrated, FPDS-NG data can help generate many of the types of insights and analyses related to cost control and budget planning that Congress seeks through the ICS. These data allow the analysis of the types of services on which DoD already spends money, the distribution of contract dollars by type of service, the number of agencies and contracting offices purchasing these services, the number of contracts used to purchase these services, and the proportion of dollars going to small businesses. Such data not only characterize currently purchased services, but they can also help improve purchasing practices. FPDS-NG data show, for example, the types of contracts used to purchase services and the extent of competition for them.

FPDS-NG data can also help illustrate changes in service spending over time, and where spending is growing or receding. For example, while DoD's overall spending and spending on services increased sharply before peaking in 2008, FPDS-NG data point to three areas—

[10] U.S. Government Accountability Office, *Civilian and Contractor Workforces: Complete Information Needed to Assess DOD's Progress for Reductions and Associated Savings*, Washington, D.C., GAO-16-172, December 2015c.

the PSC categories Construction of Structures/Facilities, Installation of Equipment, and Support Services (Professional/Administrative/Management)—where spending on services increased most rapidly. In particular, these data can show which service expenditures likely increased because of extraordinary events, such as military action in Iraq and Afghanistan, and which areas are a source of concern when spending on services continues to rise.

Finally, FPDS-NG data, when coupled with budget projections, can offer some insights on future trends in spending on services. Current budget projections show that O&M spending, which accounts for much of DoD's spending on services, continues to decrease. Assuming the proportion of O&M spending devoted to services remains relatively constant, one can reasonably conclude that overall service spending will continue to decrease. Indeed, our analysis of FPDS-NG data suggests that DoD service spending decreased by $38 billion between FY 2012 and FY 2015, exceeding congressional goals that savings in civilian and contractor workforces be at least as large (in percentage terms) as those in military basic pay as a result of reductions in military end strength.

CHAPTER SIX

Risks and Benefits of Potential Alternative Methods of Data Analysis to Inform Congressional and Defense Decisionmaking

As mentioned in previous chapters, congressional intent underlying the ICS requirement was multifaceted: to control costs and spending on service contracts, to inform strategic workforce planning efforts (particularly insourcing decisions), and to gain more oversight of number of deployed contractors in theaters of conflict. While efforts to gain oversight of deployed contractors are best met through actual counts of contractor *personnel* on the ground in theater (rather than contractor FTEs), the ICS represents an effort to calculate contractor FTEs with the aim of informing the first two of these congressional goals. Chapter Five elaborated on the utility of existing data to provide insights that would enable better and more efficient cost control over service contract spending; this chapter, meanwhile, describes how existing data can inform insourcing decisions across DoD with at least as much accuracy as ICS data.

Indeed, while the shortcomings of the ICS limit its accuracy and, thus, its ability to inform decisions about sourcing, level of contracting, and workforce planning, one of DoD's stated motives for collecting data on the direct labor hours associated with a contract is to use the information to assess the scale of its contracted force relative to in-house activities.[1] However, if the primary motivation for collecting data on contractor direct labor hours in the ICS is to inform insourcing deci-

[1] As discussed earlier in this report, such labor comparisons of service contracts versus in-house activities are relevant only to a subset of decisions related to staff augmentation contracting. For mixed or complete contracting, uncertainty regarding variability in nonlabor inputs to production that enhance productivity would make such comparisons impossible.

sions and overall strategic workforce planning through an assessment of the relative scale of the contracted workforce compared with the in-house DoD workforce, DoD should consider alternative measures that do not require collecting, validating, auditing, and protecting the proprietary data reported by contractors. From the standpoint of a cost-benefit analysis, this makes sense: Alternative metrics to calculate contractor FTEs using existing data are at least as accurate as contractor FTEs calculated using the unverified (and often estimated) direct labor hours collected in the ICS. The alternative metrics described in this chapter can be produced at a much lower cost because they rely on existing data sources (and, hence, do not require contractors to input data, which could generate cost savings on individual contracts). The use of these alternative metrics in lieu of contractor-reported or estimated direct labor hours could also assist DoD in producing an ICS in a more timely manner, as they might not be as time-consuming to generate. This could, in turn, allow the ICS to be produced in a shorter time frame, making it more usable to DoD stakeholders responsible for acquisition, budgeting, and strategic workforce planning.

This chapter describes the preliminary development of three potential alternative metrics designed to assess the scale of the contracted workforce relative to the in-house workforce performing non-contracted services for DoD. Each of these alternative metrics offers a different way to compare the relative importance of in-house and contracted services in a DoD organization using a common scale: number of FTEs. Each alternative also provides information that could inform DoD decisions relevant to sourcing strategy without requiring proprietary information from contractors. We describe the formulation of each metric, the conceptual advantages and disadvantages of each, and how they compare with each other. We also summarize our empirical comparison of the alternative metrics and the metrics currently used in the ICS (reported and estimated numbers of FTEs). The calculations underlying this empirical comparison are described in greater detail in Appendix E.

As we sought to derive alternative metrics to assess the relative scale of the service contractor workforce compared with the in-house DoD workforce, it became clear that because so many DoD organizations use military personnel and support services (such as facilities and the services of capital assets) without paying for them from their current budgets, DoD does not have accurate dollar measures of the size of most DoD organizations. Therefore, the dollar value of a contract cannot, by itself, be used to assess the relative scale of the contracted versus in-house DoD workforces. However, DoD can and does count the number of personnel in its organizations. Therefore, a comparable count of contractor personnel can provide a measure of the scale of contracting that DoD and Congress could use to inquire as to the relative importance of contracting for services as opposed to relying on in-house activities in any given DoD organization. And, in fact, Congress has mandated the collection of data on contractor direct labor hours in a form that explicitly supports such a comparison.

The first alternative metric, "civilian labor FTEs per contract," assesses the number of government civilians who could be funded if a contract service were curtailed. In effect, it expresses the size of a contract in terms of an "equivalent" count of government civilians to allow a direct comparison of the scale of work done in-house and via contract in a given DoD organization.

The second alternative metric, "contractor labor FTEs per contract," identifies the type of work conducted under a contract, the labor cost in the relevant industry, and the number of FTEs a contract of the specified size could afford to pay. This yields an alternative count of contractor labor that can then be compared with the count of government civilian or military labor in the DoD organization that funds the contract.

The third alternative metric, "contract employees as a proportion of overall contractor revenue," measures the revenue generated by a contract with a DoD organization as a fraction of the contractor's total revenue. It assumes that the contractor labor associated with the contract will be the same fraction of the contractor's total labor force. This yields a third count of contractor labor that can then be compared with the count of government civilian or military labor in the DoD organization that funds the contract.

Description and Implementation of Current and Proposed Alternative Metrics for Deriving Estimates of Contractor Full-Time Equivalents

Table 6.1 compares our three alternative metrics with reported and estimated FTEs, the metrics currently included in the ICS.[2] This illustrative exercise uses Army service contracts because reported direct labor hours were available for all such contracts. The table describes the current and alternative metrics for the R425 (Support—Professional: Engineering/Technical) and Q201 (Medical—General Health Care) PSCs. R425 includes systems engineering, technical assistance, and other services used to support a system program office during the acquisition cycle. The Army ICS data indicate that this PSC covers a significant percentage and dollar amount of Army contracted services. Q201 offers a different perspective because, while the dollar amount of the contracts is large, the number of contracts involved is substantially smaller than for R425.

The bottom section of the table provides summary statistics (average, median, and standard deviation) for the contract values, reported FTEs, estimated FTEs, civilian labor FTEs per contract, contractor labor FTEs per contract, and contract employees as a proportion of overall contractor revenue, using Army contracts in the two aforementioned PSC codes and the additional PSC codes D399 (IT and Telecom—Other IT and Telecommunications), J015 (Maintenance/Repair/Rebuild of Equipment—Aircraft and Airframe Structural Components), and S216 (Housekeeping—Facilities Operations Support). Appendix E provides a more detailed explanation of our analysis.

The two metrics currently used in the ICS are summarized in the second and third columns of Table 6.1. Under current guidance, the preferred metric is the count of direct labor hours that contractors report through the four different instances of the CMRA. Dividing this count by the number of hours in a year, 2,080, translates this count into a measure of contractor FTEs. As discussed earlier, the Army has been the most successful DoD component in collecting such data.

---

[2] ICS data for FY 2015 were not available at the time this research was conducted.

Table 6.1
Current and Proposed Metrics to Assess Contractor Manpower Numbers

			Current Metrics (median)		Proposed Alternative Metrics (median)		
PSC	Contract Amount Group	Median Contract Amount ($)	Reported FTEs	Estimated FTEs	Civilian Labor FTEs per Contract	Contractor Labor FTEs per Contract	Contract Employees as a Proportion of Overall Contractor Revenue
R425	Low	188,487	1	6	13	9	6
	Middle	2,827,120	13	13	29	21	6
	High	36,913,209	123	459	1,011	709	2,057
Q201	Low	64,039	1	5	8	7	13
	Middle	464,931	4	10	18	18	21
	High	2,508,049	13	52	89	87	127
All contracts (PSCs R425, Q201, D399, J015, and S216)	Average	11,816,441	57	184	334	263	23,887
	Median	1,216,502	7	11	21	18	14
	Standard deviation	48,391,380	223	1,623	2,799	1,954	371,749

The second metric currently used in the ICS applies a factor to the value of a contract to estimate the number of contractor FTEs associated with the contract. Today, this factor is based on data collected by the Army. For each sector defined by PSCs, the Army counts the number of FTEs and the total value of contracts for each PSC sector. It then divides the count of FTEs by the value of the contracts to yield a factor for each PSC sector. Multiplying this factor by the value of the contracts in a sector yields an estimate of the number of contractor FTEs in the sector. This metric is the dominant approach used outside the Army, but, as noted in previous chapters of this report, the rest of DoD is trying to move toward requiring reported FTEs for all contracts rather than a selected subset.

The first alternative to the metrics currently used in the ICS, "civilian labor FTEs per contract," measures the average cost of a government civilian in a DoD organization. It then divides the dollar value of a contract by this figure to determine the number of government civilians this organization could fund if it (1) eliminated the contracted services and (2) maintained its current grade plate—that is, its current mix of government civilian grades. The average cost of a government civilian can be measured in many ways. Perhaps the most reliable way is to collect data from the organization's general ledger on the dollars applied to civilian labor and the FTEs of civilian labor during a particular period. Each military service maintains its own general ledger. For example, the General Fund Enterprise Business System (GFEBS) could provide this information for the Army, and the Commander's Resource Integration System could provide this information for the Air Force. DoD could draw data from such information systems automatically through new links between these ledgers and the ICS, or it could input the data by hand if that were more cost-effective. An automated system would likely reduce the error rate that inevitably accompanies manual data transfer.

The second alternative, the "contractor labor FTEs per contract" metric, determines the contractor's industry and the location where the contracted service is provided. It uses existing federal government data to determine the per-person cost of labor in the industry and location, and it then divides the dollar value of the contract by this per-

person cost. The result is a measure of the number of personnel the contract could fund if all of its value were used to pay for personnel. Each quarter, the Bureau of Labor Statistics fields the Quarterly Census of Employment and Wages (QCEW), which covers 98 percent of U.S. jobs. QCEW data are available at the county, metropolitan statistical area, state, and national levels by industry, defined by a six-digit NAICS code. Nearly every six-digit NAICS industry is covered. The census provides detailed industry data on employment, hours, and earnings. The data are publicly available via the Bureau of Labor Statistics website and include average annual incomes within industries whose constituents are employees of DoD contractors. The result is an estimate of contractor employee earnings.

The third alternative, the "contract employees as a proportion of overall contractor revenue" metric, uses existing, publicly available data from SAM on the average value of a contractor's total annual revenues and average number of employees. Federal registration, required for all firms seeking to do business with the federal government, records the contractor's reported average annual revenue and number of employees over the past three years. It combines this information with publicly available data on the value of the firm's contracts with DoD at various work locations. It assumes that ratio of revenue to employees is the same in all the contractor's activities. If that is appropriate, dividing the dollar value of a contract by this ratio yields an estimate of the number of employees the contractor uses to execute the contract.

The "civilian labor FTEs per contract" metric uses the average cost of a government civilian to translate the dollar value of a contract into a personnel count that DoD and Congress can use to compare the scale of contract and in-house activities in any DoD organization. The "contractor labor FTEs per contract" metric uses the average cost of a contractor employee to translate the dollar value of a contract into a corresponding personnel count. Similarly, the "contract employees as a proportion of overall contractor revenue" metric produces a count of the number of contractor employees that DoD and Congress can potentially use to assess the relative importance of contract and in-house activities in any DoD organization.

Advantages and Disadvantages of the Five Metrics

The current ICS metrics, reported FTEs and estimated FTEs, have advantages and disadvantages, as do the three alternative metrics that we propose. These advantages and disadvantages are both conceptual and empirical. We first discuss the conceptual issues and summarize our comparative analysis of data on five Army service contracts to provide empirical insights. (A more detailed description of this analysis can be found in Appendix E.)

Conceptual Advantages and Disadvantages

Conceptually, the main advantage that each of the three alternatives share is that they do not require DoD to collect, validate, audit, and then protect proprietary data from DoD contractors. The high variance in contractor-reported direct labor hours in CMRA calls into question the quality and accuracy of these unverified data. Collecting, validating, auditing, and protecting DoD contractors' proprietary data is costly. Under current ICS arrangements, DoD bears this cost, and an accidental release of proprietary data can impose additional costs on contractors. Each alternative metric described here uses either existing, in-house federal data or publicly available data. If DoD used any or all of these sources, it would incur the costs of drawing data from existing sources for use in the ICS, but these costs would not differ much from current costs. Relative to reported FTEs, the main disadvantage of the estimated FTE metric currently used in ICS and our three alternatives is that they are proxies, or estimates whose usefulness depends on the assumptions required to calculate and apply them in a policy setting. Among the assumptions is that the reported FTEs are correct despite not being verified, and that they are equally applicable to all contractors in an industry, despite a large degree of variance within each industry.

The "civilian labor FTEs per contract" metric could be produced at minimal cost by merging the ICS (or FPDS-NG databases containing contract amounts) with the general ledgers of the DoD components. These existing data sources maintain their own validation and auditing programs and offer historical data that DoD could use to assess trends.

The "contractor labor FTEs per contract" metric can be produced at minimal cost once data from the ICS/FPDS-NG databases have been aligned with those from the Bureau of Labor Statistics QCEW database. DoD can use information on contractor location (by state, county, and industry) and NAICS six-digit code to achieve this alignment. This metric implicitly assumes that the QCEW data are well suited to DoD contractor workforces in the locations and industries used to align the databases. The alignment will never be perfect because NAICS codes can be broad, and the match of NAICS codes to individual contracts is imprecise. However, all the data are public and easily obtained, and historical data are available to facilitate analyses of past trends. The Bureau of Labor Statistics also performs audits to verify QCEW data quality.

The third alternative metric ("contract employees as a proportion of overall contractor revenue") relies on a combination of data from the FPDS-NG and SAM databases. As noted earlier, these existing data sources maintain their own validation and auditing programs and offer historical data that DoD could use to support trend analysis. DoD can also access these databases at low cost. This metric is most useful when it is realistic to assume that the labor intensity of a contractor's work for DoD is the same as the labor intensity of its other work. Each of these advantages and disadvantages is shown in Table 6.2.

Summary of Empirical Illustrations of the Metrics

In Table 6.3, we describe the empirical calculation of each of the current and proposed metrics, summarizing the previous descriptions. To reiterate, the methodologies employed in the three proposed alternative metrics were as follows:

- The "civilian labor FTEs per contract" metric is the ratio of the contract value to the weighted average of partially burdened annual civilian earnings. In other words, this metric divides the contract value by the average cost of a government civilian in a DoD organization.
- The "contractor labor FTEs per contract" metric is the ratio of the contract value to local/state and NAICS industry average private-

Table 6.2
Conceptual Advantages and Disadvantages of Various Metrics for Estimating Contractor Manpower

	Current Metrics		Proposed Alternative Metrics		
Advantage	**Reported FTEs**	**Estimated FTEs**	**Civilian Labor FTEs per Contract**	**Contractor Labor FTEs per Contract**	**Contract Employees as a Proportion of Overall Contractor Revenue**
Contractor-reported	X				
Costless availability using existing DoD systems and data		X	X		X
Available prior contract award (for trend analysis)		X	X	X	X
Considers contractor labor costs	X			X	
Auditable at minimal cost			X	X	X
Frequently updated data			X	X	
Disadvantage					
Requires new system or new data collection	X		X	X	
Assesses scale rather than employee-level comparisons		X	X	X	X
Requires matching by imprecise industry classifications		X	X	X	
Industry/sector- or location-level aggregation		X		X	

Table 6.3
Methods Employed by Current and Proposed Alternative Metrics to Calculate Contractor FTEs

	Current Metrics		Proposed Alternative Metrics		
Method	**Reported FTEs**	**Estimated FTEs**	**Civilian Labor FTEs per Contract**	**Contractor Labor FTEs per Contract**	**Contract Employees as a Proportion of Overall Contractor Revenue FTE**
Description of inputs for each metric	Observed direct labor hours; number of hours per year worked by FTE labor (2,088)	Contract amount ($); contract factor (industry average of FTEs ÷ contract amount [$])	Contract amount ($); weighted average of partially burdened annual civilian earnings	Contract amount ($); state and NAICS industry average of private-sector earnings	Contract amount ($); average contractor revenue; average contractor number of employees
Calculating FTEs	Observed direct labor hours ÷ 2,088	Contract amount ($) × contract factor	Contract amount ($) ÷ weighted average of partially burdened annual civilian earnings	Contract amount ($) ÷ average private-sector earnings	(Contract amount [$] ÷ average contractor revenue) x average number of employees

sector earnings. To put it differently, this metric divides the contract value by the average cost of a private-sector employee in the same industry and location as the contractor.
- The "contract employees as a proportion of overall contractor revenue" metric calculates the ratio of the contract value to the contractor's average annual revenue, multiplied by the contractor's average number of employees. This metric therefore identifies the proportion of the contractor's overall revenue that accrues from the specific contract in question and assumes an even per-dollar balance of employees across contracts to estimate the number of contractor employees dedicated to working on the contract.

Overall, estimates generated by the "contractor labor FTEs per contract" metric are very similar to those produced by the "civilian labor FTEs per contract" metric, and both are less similar to the estimated FTEs and the reported FTEs. The third alternative metric, "contract employees as a proportion of overall contractor revenue," produces the largest amount of variance. Some of its FTE estimates exceed reported FTEs, while others do not. The difference between the estimated FTEs generated by this metric is greatest for high-value contracts, and—similar to the other alternative metrics—this is likely due to FTE underestimation or significant nonlabor costs.

Comparison

Each of the metrics—both those currently in use and the three alternatives we propose—has limitations. Each can assist in projecting the scale of the workforce but does not inform capabilities and future needs. They conflate labor and nonlabor costs and, consequently, cannot separate those inputs for projections and sourcing decisions. The alternative metrics are relatively comparable to the reported FTEs, particularly for low contract values.

Correlations between the various metrics for the two PSCs we selected for our comparison exercise (R425 and Q201), as well as between the various metrics for the full sample of contracts from three additional PSCs (D399, J015, and S216), can be found in Appendix E. Correlations inform our goal of finding alternative ways to measure the scale of contracting relative to in-house activity across DoD. In general, the metrics are highly correlated. The level of correlation does vary slightly by PSC, and "contract employees as a proportion of overall contractor revenue" is the metric that deviates the most from the other metrics, likely because of the larger number of inputs to the metric and the variation associated with each input. Nonetheless, that metric has the advantage of using SAM data that contractors already provide, whereas data for the other metrics have to be tracked down.

The analyses summarized here and described more fully in Appendix E suggest that the proposed alternative metrics (or other potential metrics using similar methodologies), could, if properly applied, provide reliable proxies for the scale of contractor activity. These proxies could guide discussions of sourcing policy in DoD and Congress, and they could be used at a fraction of the cost of collecting, validating, auditing, and protecting proprietary contractor-reported FTEs.

CHAPTER SEVEN

Conclusions

Our findings suggest that the ICS products, and the processes used to create them, are not currently meeting either congressional or DoD stakeholder needs. Several factors led us to this conclusion.

First, the congressional intent underlying the ICS requirement is multifaceted and not always clearly specified in statute.

Second, ICS stakeholders are based in a range of distinct functional communities, each with its own needs and purposes for the ICS data—and these needs and purposes do not always align across communities.

Third, direct labor hours and nonlabor costs are the only unique data elements in the ICS, and there is significant disparity in opinions regarding both the quality and utility of direct labor hours among stakeholders both within and outside of DoD. For example, direct labor hours do not account for differences in productivity. This is problematic, given the variety of service contracting that DoD undertakes. Ultimately, due to the ICS's focus on direct labor hours and delayed availability, the system's data do not support timely spend analyses trends, forecasting, or strategic sourcing.

Fourth, the majority of ICS data through FY 2014 (the most recent year for which ICS data were available at the time of this research) were estimated using algorithms developed by the Army. These estimates are based on unverified contractor-reported data that vary widely in terms of numbers of reported personnel for the value of a given Army service contract in a given PSC. Thus, their validity is questionable from the outset—particularly for contracts held by a DoD component other than the Army. Some stakeholders hoped that the quality of direct labor hours data in the forthcoming eCMRA would improve over time as more contractors reported actual direct labor hours (to replace the current high levels of estimated direct labor hours). However, others feared that even the validity of actual contractor-reported direct labor hours would inevitably be questionable due to the inability to verifiably audit or benchmark them against a standard, precluding informed decisions about sourcing and budgets.

Moreover, the ICS data do not currently support spend analyses, trend analyses, forecasting, or strategic sourcing decisions, and more information would be needed to conduct effective labor comparisons to inform insourcing decisions. Other sources of data may provide many of the insights that Congress seeks to gain through efforts such as the ICS. Data from the FPDS-NG, for example, can yield insights on the specific types of services that DoD purchases, the means it uses to purchase services, trends in spending, and opportunities for increasing its buying leverage in services. Coupled with budget projections, FPDS-NG data can also indicate likely future trends in service spending and how they may compare with changes in other budget categories, such as personnel.

Other sources of data can provide different methods for estimating and reporting contractor personnel use. Alternatives that use available, in-house federal data or publicly available data would not require DoD to collect, validate, audit, and protect contractors' proprietary data. Our research indicates that internal data can yield close proxies for ICS metrics.

These findings led to several recommendations. First, policymakers should institutionalize the development and reporting of DoD-wide spend analyses of services, including analyses of trends, forecasts, and FTEs. This would entail issuing a detailed requirement for an institutionalized capability to analyze data on DoD service contracts and providing the necessary funding for its development. DoD would also likely need to employ dedicated research programmers or statistical analysts in long-term positions to produce ICS-related analyses.

Second, ICS-related statutory requirements could be refined to better distinguish between different types of contracting and, accordingly, to require the collection of different data elements for each. Our research found that DoD contracting practices vary with both the types of services purchased and the level of oversight DoD expects to give such purchases. ICS requirements could be revised to identify and distinguish among staff augmentation, mixed contracting, and complete contracting arrangements. This would likely require DoD to develop a clear definition of each of these three types of contracts, which could then be used by contracting officers to categorize contracts when initially writing them. The criteria outlined in Table 4.1 in Chapter Four—specifically, that staff augmentation contracts can be identified by the host firm (DoD, in this case) providing all inputs to a product other than some portion of the labor, by a contractor that provides only some portion of the labor, and by a predominant requirement for personnel management rather than management of costs and performance—should be instructive in creating this definition. For staff augmentation contracts, ICS requirements could be revised to specify the use of multiple alternative metrics relying on existing data sources, such as the FPDS-NG, to estimate a likely range of contractor FTEs. For cases of mixed and complete contracting, the ICS requirement could be rewritten to focus on measuring total cost and performance, rather than direct labor hours. Finally, for operational support contracts—for which congressional goals are to increase oversight of the numbers of deployed contractors on the ground—reporting requirements should focus on the number of actual deployed contractor personnel, not FTEs.

Third, DoD should periodically perform sourcing analyses of selected commercial services to determine whether civilians or contractors deliver the required level of performance at the lowest total cost. Doing so will ensure continuous adjustment of task assignments across the total force, where necessary, to maintain the lowest-cost and most effective staffing solutions for a diverse set of defense functions.

APPENDIX A

Spending on Services in the U.S. Economy and U.S. Department of Defense Over Time

Since 1947, the Bureau of Economic Analysis has published annual data on GDP, showing estimates of value added by industry. It defines an industry's value added as its gross output (e.g., sales or receipts) less its intermediate inputs (e.g., raw materials).[1]

We used these estimates to show the service sector's share of the overall economy. We define *services* as comprising the following sectors:

- utilities
- wholesale trade
- retail trade
- transportation and warehousing
- information
- finance, insurance, real estate, rentals, and leasing
- professional and business services
- educational services, health care, and social assistance
- arts, entertainment, recreation, accommodations, and food services
- other services, except government.

Since 1948, DoD has published annual data on its total obligational authority by public law title.[2] Subtracting its total obligational authority for military and civilian personnel from the overall amount

[1] Bureau of Economic Analysis, "Industry Economics Account Information Guide," web page, last updated October 16, 2016b.

[2] See, for example, Office of the Under Secretary of Defense, Comptroller, 2016.

allowed us to estimate DoD's total expenditures for goods and services. For example, in FY 2015, DoD obliged $565.4 billion, of which $139 billion was for military personnel and $67.8 billion was for civilian pay, leaving $358.6 billion for purchases of goods and services.

Figure A.1 shows how services as a share of GDP and the purchase of goods and services by DoD have grown over time. Services, which accounted for only 47 percent of GDP in 1947, accounted for 68 percent in 2015. Purchases of goods and services, which accounted for 28 percent of DoD's budget in FY 1948, accounted for 64 percent in FY 2015.

Growth in services across the economy has been concentrated in relatively few sectors, as shown in Figure A.2. Just three industry sectors—finance, insurance, real estate, rentals, and leasing; profes-

Figure A.1
Services as a Share of the U.S. Economy and Purchases as a Percentage of DoD's Budget

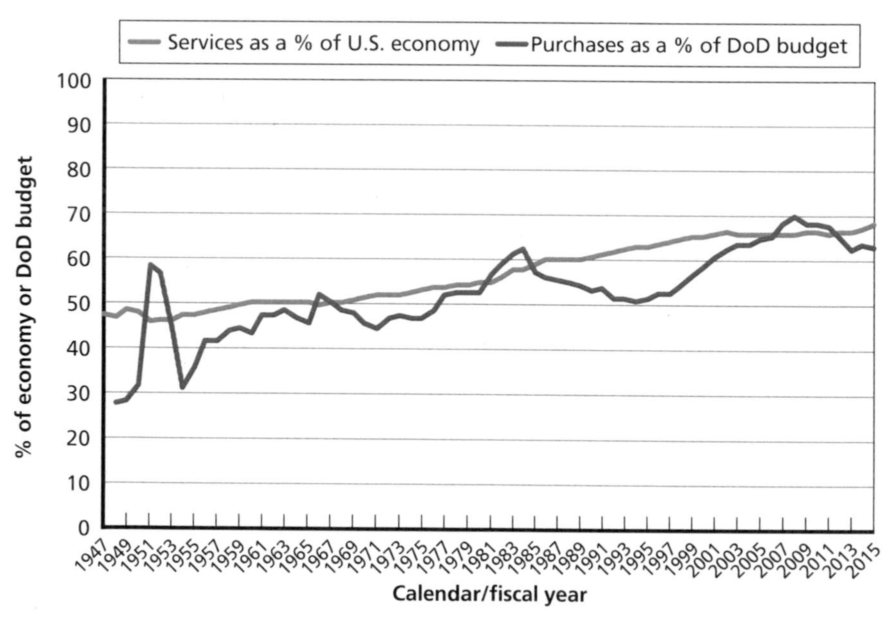

SOURCE: Bureau of Economic Analysis, 2016a; Office of the Under Secretary of Defense, Comptroller, 2016.

RAND *RR1704-A.1*

Figure A.2
Growth of Services, by Sector

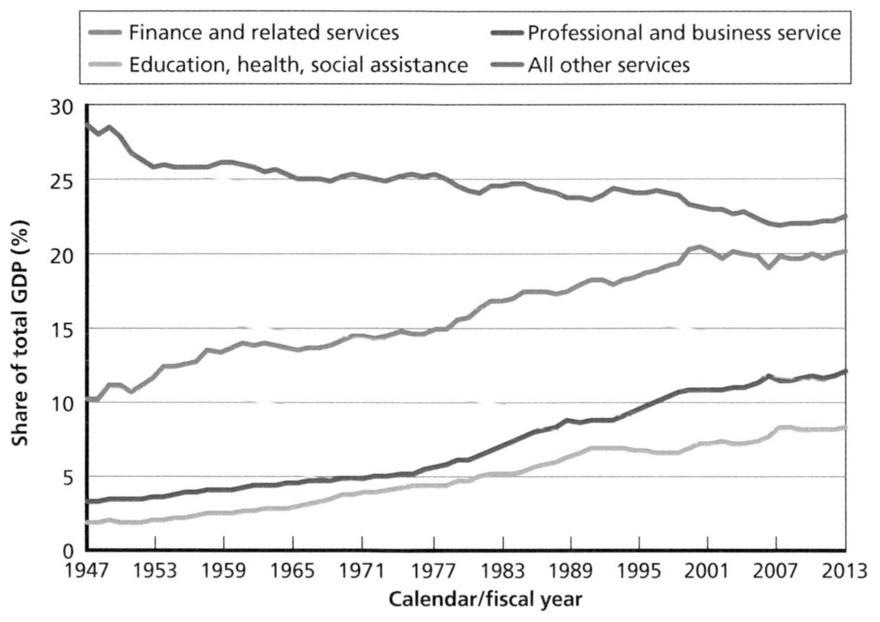

SOURCE: Bureau of Economic Analysis, 2016a.
RAND *RR1704-A.2*

sional and business services; and educational services, health care, and social assistance—more than account for all the growth in the share of services in the U.S. economy since 1947. These three sectors combined grew from 15 percent of the economy in 1947 to 41 percent in 2015. All other services combined decreased as a share of the economy, from 29 percent in 1947 to 23 percent in 2015. These trends suggest that the services that are becoming more important in the overall economy are becoming more important to DoD as well.

DoD spending on goods and services can be broadly portrayed by public law title. For example, Figure A.1 showed DoD spending on goods and services as a residual category of the total obligational authority less those for military and civilian personnel. Figure A.3 shows this spending in more detail, particularly the growth in "other goods and services." This residual category reflects the total obligational authority less those for military personnel and civilian personnel as well as less those for procurement (i.e., purchase of weapon systems), military construction, and RDT&E. The two personnel categories each account for about half the share of DoD's budget that they did decades ago. Among goods and services, procurement of weapon systems has remained within a broad band of 15–25 percent of DoD's budget in recent decades, while RDT&E remained between 10 and 14 percent. Military construction has remained at a steady but small portion of the budget, but all other goods and services have increased sharply, from less than 10 percent to more than 30 percent. While other goods and services purchases ranked below traditional military personnel, civilian personnel, and procurement spending in past decades, they now rank above these categories.

Figure A.3
DoD Purchases Relative to Personnel Spending

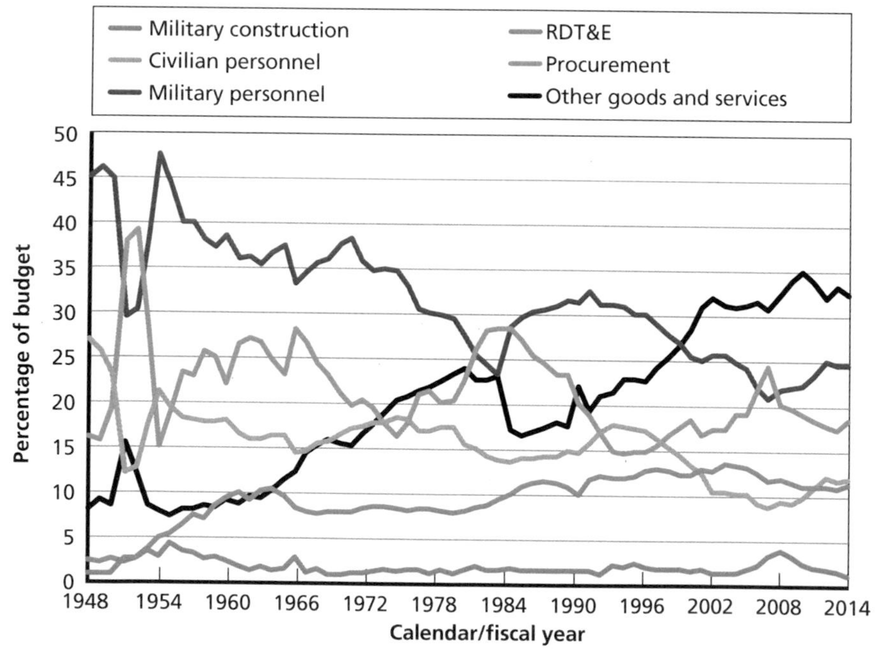

SOURCE: Office of the Under Secretary of Defense, Comptroller, 2016.
RAND *RR1704-A.3*

APPENDIX B

Summary of FY 2014 ICS Report

Table B.1 summarizes the total number of contractor FTEs and the total value of contracts, by defense organization, as reported in the ICS for FY 2014.

Table B.1
Summary of the FY 2014 ICS Report

Reporting Component	Derived or eCMRA-Reported Total Contractor FTEs	Total Obligated or Invoiced Amount ($ millions)
Department of the Navy	236,762	38,185
Department of the Army	185,654	34,827
Department of the Air Force	123,668	23,015
Defense Health Affairs	14,778	18,623
Defense Information Systems Agency	13,134	2,539
Missile Defense Agency	12,380	5,175
Defense Logistics Agency	11,668	1,706
U.S. Special Operations Command	8,548	1,483
Defense Commissary Agency	6,445	407
Defense Threat Reduction Agency	5,379	1,013
Office of the Secretary of Defense	5,202	1,063
Washington Headquarters Service	2,233	377

Table B.1—Continued

Reporting Component	Derived or eCMRA-Reported Total Contractor FTEs	Total Obligated or Invoiced Amount ($ millions)
Joint Chiefs of Staff	2,227	470
Defense Human Resource Activity	1,949	335
U.S. Southern Command	1,910	263
DoD Education Activity	1,827	225
U.S. Strategic Command	1,448	236
U.S. Transportation Command	1,000	181
U.S. Central Command	610	100
Pentagon Force Protection Agency	607	89
Defense Advanced Research Projects Agency	472	75
Defense Contract Management Agency	430	72
Defense Finance and Accounting Service	382	71
Defense Security Service	355	80
Defense Micro-Electronics Activity	322	32
Defense Legal Services Agency	273	39
Defense Acquisition University	261	59
U.S. Northern Command	254	40
U.S. European Command	252	52
Defense Security Cooperation Agency	239	45
Uniformed Services University of the Health Sciences	183	26
Defense Media Activity	127	21
Defense Technical Information Center	123	28
DoD Inspector General's Office	103	31

Table B.1—Continued

Reporting Component	Derived or eCMRA-Reported Total Contractor FTEs	Total Obligated or Invoiced Amount ($ millions)
U.S. Forces Korea	100	19
National Defense University	77	13
Defense Contract Audit Agency	45	9
Defense Intelligence Agency[a]	—	—
National Reconnaissance Office[a]	—	—
U.S. Africa Command[a]	—	—
U.S. Pacific Command[a]	—	—
Total	641,428	130,642

SOURCE: Office of the Under Secretary of Defense for Acquisition, Technology, and Logistics, *Report to Congress: Fiscal Year 2014 Inventory of Contracted Services*, August 2015.

[a] Information available in a classified annex. These data are not included in the totals.

APPENDIX C

Completeness and Quality of the ICS Data

We analyzed Army ICS data and compared them to FPDS-NG data to get a sense of the quality of the contractor-reported data.

Given that the ICS does not include all service contracts, we first compared Army ICS data, which are 100-percent contractor-reported, with Army FPDS-NG contract action data. Table C.1 shows this comparison and also includes DoD FPDS-NG contract action data. As noted in this report, ICS legislation excludes the reporting of contracts less than $150,000, as well as R&D and military construction contracts. However, DoD includes some R&D in its ICS reporting. The difference in total service contract spending between the Army FPDS-NG data and the ICS data was more than $14 billion, or almost 30 percent of Army FPDS-NG spending in 2014. This high variance is an indication of the incompleteness of the ICS data for services.

When we tried to match Army ICS data to FPDS-NG data, we could not find 1,847 contract numbers in the Army FY 2014 ICS data. These contracts represented $7,198 million that could not be matched to FY 2014 FPDS data (20.7 percent of dollars reported in the FY 2014 Army ICS data). We also found that four PSCs in the Army FY 2014 ICS data (AJ20, R&D—Math and Computer Sciences; AJ30, R&D–Environmental Sciences; AJ40, R&D—Engineering; AJ50, R&D—Life Sciences) did not exist on any available PSC lists (2007, 2009, 2011, and 2015). They appeared on 19 ICS contracts worth $39.3 million. This raises concerns about the quality and completeness of the ICS data.

Table C.1
Comparing FPDS-NG and ICS Contract Spending Reports

	DoD		Army			
	FPDS-NG		FPDS-NG		ICS[a]	
Contract Value	$ millions	%	$ millions	%	$ millions	%
< $0[b]	–3,470	–2	–1,993	–4	0	0
Less than $150,000	2,084	1	828	2	192	1
More than $150,000	157,761	101	50,306	102	34,634	99
Total	156,375	100	49,140	100	34,826	100

SOURCES: Army FY 2014 ICS data and FY 2014 FPDS-NG awards (data as of March 2015).

[a] Excludes contracts less than $150,000 and R&D and military construction contracts.

[b] Indicates de-obligations, corrections, or modifications.

Given that staff augmentation contracting is most likely to be found in support services, we analyzed Army data for PSC category R, Support (Professional/Administrative/Management), overall, as well as three individual PSCs in this category:

- R408, Support—Professional: Program Management/Support
- R425, Support—Professional: Engineering/Technical
- R499, Support—Professional: Other.

In analyzing these data, we found very high variance between the minimum and maximum, as well as between the median and average, and a high standard deviation of contract dollars per contractor-reported FTE, as shown in Table C.2. This raises concerns about the quality of the reported direct labor hours data.

We next took the top 11 PSCs by total dollars in the Army ICS and calculated the minimum, quartile 1, median, quartile 3, and maximum dollars per FTE for each, as shown in Table C.3. Again, we see considerable to significant variation in the values.

Taken together, the incompleteness of ICS data coupled with high to very high variance in reported dollars per FTE raises serious questions about the quality and value of ICS data, particularly for workforce planning.

Table C.2
Variance in Minimum, Maximum, Median, Average, and Standard Deviation of Dollars per FTE, by PSC

	$ per FTE			
Statistics	**PSC Category R Support (Professional/ Administrative/ Management)**	**R408 Support—Professional: Program Management/ Support**	**R425 Support—Professional: Engineering/ Technical**	**R499 Support—Professional: Other**
Minimum	3,194	29,232	15,993	8,686
Median	173,549	196,348	222,205	166,002
Average	719,723	835,908	502,662	494,010
Maximum	323,677,584	226,232,712	21,301,860	68,321,866
Standard deviation	8,874,570	9,756,322	1,554,060	3,289,628

SOURCES: Army FY 2014 ICS data and FY 2014 FPDS-NG awards (data as of March 2015).

Table C.3
Variance in Minimum, Quartile 1, Median, Quartile 3, and Maximum Dollars per FTE in Top 11 PSCs Raises Questions About Army ICS Data Quality and Value

	$ per FTE				
Top PSC in ICS	Minimum	Quartile 1	Median	Quartile 3	Maximum
R408, Support—Professional: Program Management/Support	29,000	130,000	196,000	290,000	226,233,000
S216, Housekeeping—Facilities Operations Support	11,000	64,000	91,000	169,000	1,540,000
J016, Maintenance, Repair, and Rebuilding of Equipment—Aircraft Components and Accessories	47,000	92,000	170,000	428,000	8,607,000
D316, IT and Telecom—Telecommunications Network Management	73,000	108,000	164,000	244,000	1,391,000
Q999, Medical—Other	40,000	85,000	136,000	312,000	105,895,000

SOURCE: Army FY 2014 ICS data as of March 7, 2014.

APPENDIX D

Interview Protocols

As noted in Chapters One and Three, this research included interviews of stakeholders across DoD, congressional staff, service contractors, non-DoD federal agencies, and other subject-matter experts from policy research organizations. These interviews used a semistructured format, guided by specific interview protocols targeted toward the population of interest. This appendix presents the four distinct protocols used for these interviews. As is typical of a semistructured interview methodology, not all questions on a given protocol were relevant to a particular interviewee. The questions were used to provide a basic structure to the conversation, but other points were often covered as well during these conversations.

Each interview included a short summary of the project at the outset of the conversation, as well as a consent script modeled on human subjects protection guidelines to indicate that all interviews were confidential and that any questions could be skipped if the interviewee was at all uncomfortable answering them.

Interview Protocol for Congressional Staff

Background

1. Please tell us about your career leading up to your current position. We are particularly interested in any experiences you may have had with workforce and/or contractor management.

Congressional Interest/Intent

1. Please describe the drivers of congressional interest in better visibility of service contracts (e.g., growth of spending, appropriateness of contracting [i.e., inherently governmental or close to inherently governmental], improved strategic workforce planning, more agile use of the total force, skirting compensation caps, unfair to federal personnel).

2. What are the goals of improved oversight of service contracts? (e.g., lowering total costs, ensuring best value, providing access to needed technical personnel, insourcing/outsourcing some work, ensuring that work is not inherently governmental or close to it)

3. Are there particular types of service contracts that are of more or less concern to Congress (e.g., specific industries, such as IT; specific types of contracts, such as time and materials contracts, labor substitution contracts; specific types of uses, such as complete outsourcing)?

4. Are there particularly types of information on service contracts Congress wants to know (e.g., direct labor only, total labor [including overhead], average cost per employee, total costs, performance metrics, such as quality, on-time delivery, unit price, and price comparisons for similar services)?

5. What would Congress like DoD to do with the information it collects on service contracts?

Interview Protocol for Defense Officials

Background

1. Please tell us about your career leading up to your current position. We are particularly interested in any experiences you may have had with personnel or contractor management or the annual Inventory for Contracted Services (ICS).

Experience with the ICS

Process

2. Please describe your experience with the ICS.
   a. When and how did you learn about the ICS?
   b. How much time does it take to collect and process these data?
   c. How many staff typically work on this task?
   d. Have there been any challenges with collecting these data (e.g., have contractors cooperated or complained, does it require special staff to field the survey and process responses)?
   e. How much has the ICS requirement increased the workload for your office/organization?

3. Have you used the eCMRA IT system to collect and report ICS data? If so,
   a. How easy/simple is it to use?
   b. What data does it collect that you do not already have?
   c. Have you received any feedback from contractors or personnel on the ease of use?
   d. Are there additional data elements that you do not have that you would like it to collect?

Substance

4. Have you used the eCRMA/ICS data for any purpose other than required ICS reporting? If yes, for what have you used it? If not, why not? (e.g., need additional data to make it useful, such

as . . . ; data not relevant to any planning or policy decisions; lack of personnel to process the data; lack of access to additional data; lack of available additional data; if so, what other data do you need?).

5. Have your checked the ICS data for quality? If so, what did you find? If not, why not?

Opportunities for Improvement

6. In your experience, what other types of data related to service contracts would be useful to improve DoD strategic planning and decisionmaking efforts to
   a. inform budget processes?
   b. inform human capital planning/strategic workforce management?
   c. determine whether contractors are performing inherently governmental functions?
   d. prevent fraud, waste, and abuse?
   e. control spending on service contracts?

Strategic Workforce Planning

7. Has your office insourced any contractor positions (i.e., gotten the billets and appropriations to hire civilian or military personnel to fill them), or does it have any plans to do so? If not, why not?

8. In practice, how does your office/organization ensure that service contractors do not perform inherently governmental functions?

Wrapping Up

9. In your experience, are there any aspects of the ICS data collection and analysis that could be improved—procedurally, substantively, or otherwise?

Interview Protocol for Non-DoD Federal (Interagency) Officials

Background

1. Please tell us about your career leading up to your current position. We are particularly interested in any experiences you may have had with the collection of data pertaining to federal service contracts, as well as experience managing personnel or contractors.

Strategic Workforce Planning

2. Has your office/agency insourced any contractor positions (i.e., gotten the billets and appropriations to hire civilian or military personnel to fill them), or does it have any plans to do so? If not, why not?

3. In practice, how does your office/agency ensure that service contractors do not perform inherently governmental functions?

Experience Developing and Using Service Contract Inventories

Process

4. Please describe what process your agency uses to inventory service contracts.
    a. Does it use existing data? If so, from what systems? (e.g., FPDS-NG) and the System for Award Management (SAM)?
    b. Does it collect special data through a survey of contractors? If yes,

        i. What data elements does it collect?
    c. Have you received any resistance from contractors while collecting these data?
    d. How much time and how many people do you need to produce the service contract inventory for your agency? Does the process require specially trained people to understand, execute, and/or summarize the data?

Substance

5. Has the quality of the data in your agency's service contract inventory been verified? If so, how, and were any quality issues encountered?

6. Do you use the service contract data you collect for this inventory for any other purposes? If so, for what?
    a. To inform budget processes?
    b. To inform human capital planning/strategic workforce management?
    c. To determine whether contractors are performing inherently governmental functions?

7. Does any relevant service contract data exist that your agency does *not* collect, but that would improve workforce or budget planning if it were collected? If so, what additional data could be collected?

8. How is your organization providing/estimating
    a. A description of the role the services played in achieving agency objectives?
    b. The number and work location of contractor employees, expressed as FTEs for direct labor, compensated under the contract?

IT Systems

9. Have you ever experienced any challenges entering relevant service contract data into, or working with the data found in FPDS-NG, SAM, or other IT systems that your agency uses for its service contract inventory?

10. Has your organization found any particular IT system to be most cost-effective and useful for collecting and reporting information on its service contract inventory?

Opportunities for Improvement

11. Have you encountered any challenges in compiling the service contract inventory for your agency? If so, what were they?

12. Has your agency identified any ways your service contract inventory could be done more efficiently or effectively?

Interview Protocol for Service Contractors

1. Please tell us about your career leading up to your current position. We are particularly interested in any experiences you may have had with the collection of performance data for your federal and commercial service contracts.

2. Please describe your business (sector, industry).
   a. When was your firm first created? When did it start providing the services it currently provides?
   b. Approximate percentage of revenue from the federal government, DoD, state/local governments, and the private sector?

3. Please describe your history with your different types of customers (e.g., federal, DoD, non–federal government, private sector).
   a. Over what time period?
   b. Number of submitted bids?

   c. Number of awards?
   d. Types of services provided?

4. Please describe your experience with the data collection processes related to the Annual Inventory of Contracted Services (ICS).
   a. How did you learn about the ICS?
   b. What have you observed to work well in these data collection processes? What have you observed to have not worked well?
   c. What data are you required to provide for the ICS and in what format?
   d. Did your firm provide any of this information to any party (government or otherwise) prior to the ICS requirement?
   e. How much time does it take to collect and provide these data in the format requested? How many of your staff typically work on this task?
   f. Does your firm directly enter the ICS data into the eCMRA IT system? If not, how/to whom are they reported?
   g. If you have used the eCMRA IT system before,
      i. How easy/simple is it to use?
      ii. Did you face any challenges in using it?
      iii. What data, if any, does it collect that you are already reporting elsewhere?
      iv. What data does it collect that you do not report elsewhere?
   h. How have you observed the ICS requirement to affect your firm's business workload?

5. Have you ever had to provide similar data to non-DoD customers? If so, what types of data?
   a. How does your experience with non-DoD customers differ?

6. In your experience, what types of data should DoD collect on its service contracts to improve its strategic planning and decisionmaking efforts to
   a. Inform budget processes?
   b. Inform human capital planning/strategic workforce management?
   c. Determine whether contractors are performing inherently governmental functions?
   d. Prevent fraud, waste, and abuse?
   e. Control spending on service contracts?

7. Have you ever received feedback that some of the work you do is inherently governmental or close to it?

8. From the standpoint of your firm, are there any aspects of the ICS data collection and analysis that could be improved—procedurally, substantively, or otherwise?

APPENDIX E

Case-Study Comparison of Current and Proposed Alternative Metrics for Contractor FTEs Using FPDS-NG Contract Values

In Table E.1, we illustrate the current and alternative metrics documented in Chapter Six using examples of Army contracts and relying on ICS data for the number of FTEs and FPDS-NG data for the contract values. This illustrative exercise uses Army service contracts because reported FTEs were available for all such contracts. For the purposes of demonstration, we first divide the sample of contracts by PSC code.

Table E.1 summarizes the current and alternative metrics using ICS data on Army contracts. The table describes the current (columns 4 and 5) and alternative (columns 6–8) metrics for PSCs R425 (Support—Professional: Engineering/Technical) and Q201 (Medical—General Health Care). R425 includes systems engineering, technical assistance, and other services used to support a system program office during the acquisition cycle. The Army ICS data indicate that this PSC covers a significant percentage and dollar amount of Army contracted services. Q201 offers a different perspective because, while the dollar amount of the contracts is large, the number of contracts involved is substantially smaller than for R425. The first three rows of Table E.1 correspond to the R425 PSC, and rows 4–6 correspond to the Q201 PSC.

The table provides a comparison of the current metrics (both reported (column 4) and estimated (column 5) FTEs), as well as the proposed alternative metrics (columns 6–8). Each of the proposed alternative metrics produced estimates of FTEs similar to the second of the current metrics (estimated FTEs, column 5). To compare the current and proposed alternative metrics of direct labor, we first divided each PSC into low, middle, and high contract value ranges. This allows us to describe how the similarities and differences between the estimated and reported FTEs vary by contract value. For R425, the low range was an annual value less than $1 million, the middle range was an annual value between $1 million and $10 million, and the high range was an annual value exceeding $10 million. For Q201, the low range was an annual value less than $250,000, the middle range was an annual value between $250,000 and $1 million, and the high range was an annual value exceeding $1 million. For each PSC (R425 and Q201), we identify the median contract by dollar value and show the FTEs as reported or estimated using each of the current and alternative metrics. In the bottom section of Table E.1, "All," we provide summary statistics (average, median, and standard deviation) for the contract values, reported FTEs, estimated FTEs, civilian labor FTEs per contract, contractor labor FTEs per contract, and contract employees as a proportion of overall contractor revenue, using Army contracts in the two aforementioned PSC codes and the additional PSCs D399 (IT and Telecom—Other IT and Telecommunications), J015 (Maintenance, Repair, and Rebuilding of Equipment—Aircraft and Airframe Structural Components), and S216 (Housekeeping—Facilities Operations Support).[1]

Current Metrics

The numbers of FTEs that contractors reported through CMRA are shown in the "Reported FTEs" column of Table E.1. The column "Estimated FTEs" shows the number of FTEs that the Army's contract factors would estimate if applied to the contract amount. Note that the reported FTEs and the estimated FTEs are generally very similar.

---

[1] We selected these five PSCs because they represent a significant percentage and dollar amount of contracted services. Note that there are differences between the contract values in the ICS and the FPDS-NG. We show the differences between the reported and estimated FTEs resulting from current and proposed metrics using ICS contract values. This is because of variation between the ICS and FPDS-NG in the actions corresponding to each contract (e.g., invoices versus obligations and de-obligations).

Table E.1
Comparing Current and Proposed Alternative Metrics of Contractor FTEs

			Current Metrics (median)		Proposed Alternative Metrics (median)		
PSC	Contract Amount Group	Median Contract Amount ($)	Reported FTEs	Estimated FTEs	Civilian Labor FTEs per Contract	Contractor Labor FTEs per Contract	Contract Employees, by Proportion of Overall Contractor Revenue
R425	Low	188,487	1	6	13	9	6
	Middle	2,827,120	13	13	29	21	6
	High	36,913,209	123	459	1,011	709	2,057
Q201	Low	64,039	1	5	8	7	13
	Middle	464,931	4	10	18	18	21
	High	2,508,049	13	52	89	87	127
All contracts (PSCs R425, Q201, D399, J015, and S216)	Average	11,816,441	57	184	334	263	23,887
	Median	1,216,502	7	11	21	18	14
	Standard deviation	48,391,380	223	1,623	2,799	1,954	371,749

Where there are differences, the estimated FTEs are typically larger than the reported FTEs. The few differences that arise between the reported and estimated FTEs are for high-value contracts.

There are three problems in evaluating the "correctness" of the estimated FTEs relative to the reported FTEs. First, the accuracy of the reported FTEs is unknown because they are neither auditable nor verifiable. There are examples of contracts exceeding $20 million with only one FTE reported—potentially indicating misreports or contractors with extremely high levels of nonlabor costs. Second, the contract factor used to estimate the FTEs is based on reported FTEs of Army contractors. To the extent that these Army contractors differ from other DoD contractors, the estimates will differ from actual FTEs across DoD. Third, there is a high degree of variance in reported FTEs for each range of contract values. The contract factor is an average of the ratio of reported FTEs to contract value, so the imprecision of the estimated FTEs will grow with the variance in reported FTEs. Consequently, while the medians of the reported and estimated FTEs within each range of contract values align well, the differences between the reported and estimated FTEs will increase toward the tails of the distribution of contract values.

Alternative Metric 1: Civilian Labor FTEs per Contract

The first alternative metric is descriptively titled "civilian labor FTEs per contract." The estimates for this metric are derived using the grade plate for the White Sands Missile Range (UIC W04WAA). If the Army were to calculate these metrics, it would use data on counts of and dollar expenditures on government civilians from GFEBS for each organization using the services of the six contracts. We did not have access to these data.[2]

---

[2] The derivation of the "civilian labor FTEs per contract" metric is different from the proposed metric because we did not have access to the GFEBS data. Consequently, the derivation matched each PSC to a UIC because the FY 2014 Tables of Distribution and Allowances (TDA) showed the hiring of contractors with the same PSC codes. Average total annual

To illustrate the use of this metric, we collected data from the TDA document that defines the authorized numbers of civilians by grade. We then applied cost factors from the Army Manpower Costing System for these grades to estimate the annual cost of the grade plates for each organization. We used these data to estimate the total annual cost of civilian personnel in each organization and then the average annual cost of these personnel. We used this average annual cost to estimate, within a fixed budget for each organization, how many government civilians the Army could fund if the contracts were terminated.

For the low contract amount range, the median FTEs estimated by the "civilian labor FTEs per contract" metric were slightly higher than the reported and estimated FTEs for both the R425 and Q201 PSCs. As the contract values grow, so too do the estimates of this metric and the differences between it and the reported FTEs. For the high contract amount range, the difference between this metric's estimate and the reported FTEs is larger in R425 than Q201. The differences between this metric and the reported FTEs may be due, in part, to using the TDA data instead of the GFEBS data. Alternatively, they may reflect underreporting of FTEs rather than overestimation by the metric.

There are a number of other potential explanations for the differences between the reported FTEs and the FTEs estimated by the "civilian labor FTEs per contract" metric. First, if the differences are not due to using the TDA data and the reported FTE levels are accurate, the overestimation may imply that contractors employ labor that costs significantly more per person than government civilians. Second, the differences could result from the contract functions being executed in higher-cost areas than the UIC selected for this exercise (White Sands Missile Range). Third, the contracts could have significant nonlabor costs. Fourth, it could be that contractor labor tends to cost more than government civilian labor for Army activities performed in the selected PSCs, particularly R425. The point of this metric is not to determine the per-person cost of contract employees. Rather, it is to show how

compensation was determined using general schedule pay grades weighted by the prevalence of hiring each grade at the UIC.

many government civilian bodies could be funded if the Army terminated each contract; that is, how many government civilian bodies are equivalent to each contractor employee.

Alternative Metric 2: Contractor Labor FTEs per Contract

The second alternative metric is descriptively titled "contractor labor FTEs per contract." To conduct this analysis, we merged the ICS data with data from the Bureau of Labor Statistics QCEW.

As with the "civilian labor FTEs per contract" metric, this alternative metric produces estimates that generally exceed the reported FTEs. Also similar to the "civilian labor FTEs per contract" metric estimates, the differences between the estimates using this metric and the reported FTEs grow with the contract value. Overall, estimates using the "contractor labor FTEs per contract" metric are very similar to those produced by the "civilian labor FTEs per contract" metric, and each are less similar to the current ICS metrics, estimated FTEs and reported FTEs.

The differences between the "contractor labor FTEs per contract" estimates and the reported FTEs may reflect underreporting of FTEs rather than overestimation by the metric. The differences may also reflect a higher per-person cost for contractor employees than the QCEW averages. Alternatively, there could be significant nonlabor costs associated with the contracts.

Alternative Metric 3: Contract Employees as a Proportion of Overall Contractor Revenue

The third alternative metric is "contract employees as a proportion of overall contractor revenue." As noted earlier, this metric calculates the ratio of the contract value to the contractor's average annual revenue, multiplied by the contractor's average number of employees. To calculate estimates of contractor FTEs using this metric, we matched the ICS contracts to corresponding information in the SAM database.[3]

This metric produces the largest amount of variance. Some of its FTE estimates exceed reported FTEs, while others do not. The full sample statistics at the bottom of Table E.1 show that the standard deviation is largest for the "contract employees as a proportion of overall contractor revenue" metric. The difference in the estimated FTEs generated by this metric is greatest for high-value contracts, and—as with the other alternative metrics—this is likely due to FTE underestimation or significant nonlabor costs. Nonetheless, this metric has the advantage of being based on SAM data that contractors already provide.

Comparison

Each of the metrics—both those currently in use and the three alternatives we propose—has limitations. They can assist in projecting the scale of the workforce, but they do not reveal capabilities and future needs. They conflate labor and nonlabor costs and consequently cannot separate those inputs to support projections and sourcing decisions. The alternative metrics are relatively comparable to the reported FTEs, particularly for low contract values. The similarities suggest that the alternative metrics are close proxies for one another, but the correlation of each varies by PSC.

Table E.2 shows the correlations between the various metrics for the R425 and Q201 PSCs, as well as between the full sample of contracts from the five selected PSCs. Correlations inform our goal of finding alternative ways to measure the scale of contracting relative to in-house activity across DoD. In general, the metrics are highly correlated, but the level of correlation varies slightly by PSC. Again, "contract employees as a proportion of overall contractor revenue" is

---

[3] Not all ICS contracts could be matched to the SAM database. In these instances, we could not produce estimates because we required SAM data on the contractor's average annual revenue and average number of employees.

Table E.2
Pearson Correlation Coefficients Among Metrics

		Current Metrics		Proposed Alternative Metrics		
PSC	Category	Reported FTEs	Estimated FTEs	Civilian Labor FTEs per Contract	Contractor Labor FTEs per Contract	Contract Employees as a Proportion of Overall Contractor Revenue
R425	Reported FTEs	1				
	Estimated FTEs	0.9891	1			
	Contract amount, per civilian dollar	0.9891	0.9985	1		
	Contract amount, per contractor labor	0.9873	0.9966	0.9966	1	
	Employees, by contract portion of revenue	−0.0160	−0.0158	−0.0158	−0.0171	1
Q201	Reported FTEs	1				
	Estimated FTEs	0.8283	1			
	Contract amount, per civilian dollar	0.8283	0.9972	1		
	Contract amount, per contractor labor	0.8433	0.9875	0.9875	1	
	Employees, by contract portion of revenue	0.6595	0.6352	0.6352	0.599	1

Table E.2—Continued

PSC	Category	Current Metrics		Proposed Alternative Metrics		
		Reported FTEs	Estimated FTEs	Civilian Labor FTEs per Contract	Contractor Labor FTEs per Contract	Contract Employees as a Proportion of Overall Contractor Revenue
All contracts (PSCs R425, Q201, D399, J015, and S216)	Reported FTEs	1				
	Estimated FTEs	0.9860	1			
	Contract amount, per civilian dollar	0.9731	0.9889	1		
	Contract amount, per contractor labor	0.9788	0.9918	0.9897	1	
	Employees, by contract portion of revenue	−0.0118	−0.0123	−0.0112	−0.0126	1

the metric that deviates the most from the others, likely because of the larger number of inputs to the metric and the variation associated with each input. However, this metric has the advantage of being based on SAM data that contractors already provide, whereas data for the other two metrics have to be tracked down.

APPENDIX F

Defense Component Plans for Compliance with 10 U.S.C. 2330a

Defense Advanced Research Projects Agency, letter to the House Appropriations Committee, House Armed Services Committee, Senate Appropriations Committee, and Senate Armed Services Committee on compliance plan for 10 U.S.C. 2330a, September 28, 2011.

Defense Commissary Agency, letter to the House Appropriations Committee, House Armed Services Committee, Senate Appropriations Committee, and Senate Armed Services Committee on compliance plan for 10 U.S.C. 2330a, September 26, 2011.

Defense Finance and Accounting Service, letter to the House Appropriations Committee, House Armed Services Committee, Senate Appropriations Committee, and Senate Armed Services Committee on compliance plan for 10 U.S.C. 2330a, August 31, 2011.

Defense Human Resources Activity, letter to the House Appropriations Committee, House Armed Services Committee, Senate Appropriations Committee, and Senate Armed Services Committee on compliance plan for 10 U.S.C. 2330a, September 14, 2011.

Defense Intelligence Agency, letter to the House Appropriations Committee, House Armed Services Committee, Senate Appropriations Committee, and Senate Armed Services Committee on compliance plan for 10 U.S.C. 2330a, September 29, 2011.

Defense Information Systems Agency, letter to the House Appropriations Committee, House Armed Services Committee, Senate Appropriations Committee, and Senate Armed Services Committee on compliance plan for 10 U.S.C. 2330a, September 30, 2011.

Defense Logistics Agency, letter to the House Appropriations Committee, House Armed Services Committee, Senate Appropriations Committee, and Senate Armed Services Committee on compliance plan for 10 U.S.C. 2330a, October 25, 2011.

Defense Legal Services Agency, letter to the House Appropriations Committee, House Armed Services Committee, Senate Appropriations Committee, and Senate Armed Services Committee on compliance plan for 10 U.S.C. 2330a, September 7, 2011.

Defense Media Activity, letter to the House Appropriations Committee, House Armed Services Committee, Senate Appropriations Committee, and Senate Armed Services Committee on compliance plan for 10 U.S.C. 2330a, September 29, 2011.

Defense Contract Audit Agency, letter to the House Appropriations Committee, House Armed Services Committee, Senate Appropriations Committee, and Senate Armed Services Committee on compliance plan for 10 U.S.C. 2330a, October 3, 2011.

Defense Contract Management Agency, letter to the House Appropriations Committee, House Armed Services Committee, Senate Appropriations Committee, and Senate Armed Services Committee on compliance plan for 10 U.S.C. 2330a, November 7, 2011.

Defense Prisoner of War/Missing Personnel Office, letter to the House Appropriations Committee, House Armed Services Committee, Senate Appropriations Committee, and Senate Armed Services Committee on compliance plan for 10 U.S.C. 2330a, September 27, 2011.

Defense Security Cooperation Agency, letter to the House Appropriations Committee, House Armed Services Committee, Senate Appropriations Committee, and Senate Armed Services Committee on compliance plan for 10 U.S.C. 2330a, October 7, 2011.

Defense Security Service, letter to the House Armed Services Committee, House Appropriations Committee, Senate Appropriations Committee, and Senate Armed Services Committee on compliance plan for 10 U.S.C. 2330a, September 28, 2011.

Defense Technical Information Center, memorandum to the Office of Under Secretary of Defense for Personnel and Readiness on compliance plan for 10 U.S.C. 2330a, September 26, 2011.

Defense Technology Security Administration, letter to the House Appropriations Committee, House Armed Services Committee, Senate Appropriations Committee, and Senate Armed Services Committee on compliance plan for 10 U.S.C. 2330a, September 30, 2011.

Department of the Air Force, letter to the House Appropriations Committee, House Armed Services Committee, Senate Appropriations Committee, and Senate Armed Services Committee on Air Force Compliance Plan for 10 U.S.C. 2330a, September 28, 2011.

Department of the Navy, memorandum for the Office of the Under Secretary of Defense for Personnel and Readiness on updated compliance plan for 10 U.S.C. 2330a, October 31, 2011.

National Defense University, memorandum to the House Appropriations Committee, House Armed Services Committee, Senate Appropriations Committee, and Senate Armed Services Committee on compliance plan for 10 U.S.C. 2330a, October 6, 2011.

National Geospatial-Intelligence Agency, letter to the House Appropriations Committee, House Armed Services Committee, Senate Appropriations Committee, and Senate Armed Services Committee on compliance plan for 10 U.S.C. 2330a, October 6, 2011.

National Reconnaissance Office, letter to the House Armed Services Committee on compliance plan for 10 U.S.C. 2330a.

National Security Agency, letter to the House Appropriations Committee, House Armed Services Committee, Senate Appropriations Committee, and Senate Armed Services Committee on compliance plan for 10 U.S.C. 2330a, October 31, 2011.

North American Aerospace Defense Command and U.S. Northern Command, letter to the House Appropriations Committee, House Armed Services Committee, Senate Appropriations Committee, and Senate Armed Services Committee on compliance plan for 10 U.S.C. 2330a, October 3, 2011.

Office of Economic Adjustment, letter to the House Appropriations Committee, House Armed Services Committee, Senate Appropriations Committee, and Senate Armed Services Committee on compliance plan for 10 U.S.C. 2330a, September 30, 2011.

Office of the Under Secretary of Defense for Acquisition, Technology, and Logistics, *Report to Congress: Fiscal Year 2014 Inventory of Contracted Services*, August 2015.

Office of the Under Secretary of Defense for Personnel and Readiness, "Action Memo on Congressional Notification Regarding OSD Efforts to Implement Paragraph (c) of 8108 of Public Law 112-10, the Department of Defense and Full-Year Continuing Appropriations Act," 2011.

———, "Coordination Sheet for Selected Component Reports, IAW 8108(c) of P.L. 112-10, to Congress on Plans for Documenting Full-Time Contractor Employees for the Inventory of Contracts for Services (ICS)," September 15, 2011.

———, "Coordination Sheet for Selected Component Reports, IAW 8108(c) of P.L. 112-10, to Congress on Plans for Documenting Full-Time Contractor Employees for the Inventory of Contracts for Services (ICS)," September 26, 2011.

Secretary of the Army, letter to the House Appropriations Committee, House Armed Services Committee, Senate Appropriations Committee, and Senate Armed Services Committee on Army compliance plan for 10 U.S.C. 2330a, September 26, 2011.

Test Resource Management Center, memorandum for the Under Secretary of Defense for Personnel and Readiness on compliance plan for 10 U.S.C. 2330a, October 7, 2011.

TRICARE Management Activity, letter to the House Appropriations Committee, House Armed Services Committee, Senate Appropriations Committee, Senate Appropriations Committee, and Senate Armed Services Committee on compliance plan for 10 U.S.C. 2330a, October 24, 2011.

U.S. Africa Command, letter to the Senate Appropriations Committee on U.S. Transportation Command Compliance Plan for 10 U.S.C. 2330a, September 26, 2011.

———, memorandum to the Office of Under Secretary of Defense for Personnel and Readiness on compliance plan for 10 U.S.C. 2330a, September 29, 2011.

U.S. Central Command, letter to the House Appropriations Committee, House Armed Services Committee, and Senate Appropriations Committee on compliance plan for 10 U.S.C. 2330a, October 4, 2011.

U.S. Department of Defense Education Activity, letter to the House Appropriations Committee, House Armed Services Committee, Senate Appropriations Committee, and Senate Armed Services Committee on compliance plan for 10 U.S.C. 2330a, September 30, 2011.

U.S. European Command, letter to the House Appropriations Committee, House Armed Services Committee, Senate Appropriations Committee, and Senate Armed Services Committee on compliance plan for 10 U.S.C. 2330a, October 1, 2011.

U.S. Pacific Command, memorandum on compliance plan for 10 U.S.C. 2330a.

U.S. Southern Command, memorandum to the House Appropriations Committee compliance plan for 10 U.S.C. 2330a, September 30, 2011.

U.S. Southern Command, memorandum to the Senate Appropriations Committee and Senate Armed Services Committee on compliance plan for 10 U.S.C. 2330a, September 30, 2011.

U.S. Special Operations Command, letter to the House Appropriations Committee, House Armed Services Committee, Senate Appropriations Committee, and Senate Armed Services Committee on compliance plan for 10 U.S.C. 2330a, September 29, 2011.

U.S. Strategic Command, memorandum for the Under Secretary of Defense for Personnel and Readiness on compliance plan for 10 U.S.C. 2330a, November 2, 2011.

U.S. Transportation Command, letter to the House Appropriations Committee, House Armed Services Committee, and Senate Armed Services Committee on compliance plan for 10 U.S.C. 2330a, September 26, 2011.

Washington Headquarters Services and Pentagon Force Protection Agency, memorandum for the Under Secretary of Defense for Personnel and Readiness on updated compliance plan for 10 U.S.C. 2330a, October 7, 2011.

Bibliography

Acemoglu, Daron, and David Autor, *Lectures in Labor Economics*, Cambridge, Mass.: Massachusetts Institute of Technology, 2016. As of February 3, 2017: http://economics.mit.edu/files/4689

Allen, Sandy, and Ashok Chandrashekar, "Outsourcing Services: The Contract Is Just the Beginning," *Business Horizons*, Vol. 43, No. 2, March 2000, pp. 25–34.

Allison, Graham, and Philip Zelikow, *Essence of Decision: Explaining the Cuban Missile Crisis*, 2nd ed., London: Pearson, 1999.

Amaral, Jason, Corey A. Billington, and Andy A. Tsay, "Outsourcing Production Without Losing Control," *Supply Chain Management Review*, November–December 2004, pp. 44–52. As of February 3, 2017: http://e3associates.com/sites/default/files/Outsourcing%20without%20losing%20control.pdf

Amey, Scott, "DoD Contractors Cost Nearly 3 Times More Than DoD Civilians," *Project on Government Oversight Blog*, November 30, 2012. As of February 3, 2017: http://pogoblog.typepad.com/pogo/2012/11/dod-contractors-cost-nearly-3-times-more-than-dod-civilians.html

Arrow, Kenneth J., H. B. Chenery, B. S. Minhas, and R. M. Solow, "Capital-Labor Substitution and Economic Efficiency," *Review of Economics and Statistics*, Vol. 43, No. 3, August 1961, pp. 225–250.

Asch, Beth J., *The Pay, Promotion, and Retention of High-Quality Civil Service Workers in the Department of Defense*, Santa Monica, Calif.: RAND Corporation, MR-1193-OSD, 2001. As of February 3, 2017: http://www.rand.org/pubs/monograph_reports/MR1193.html

Baker, George P., Michael C. Jensen, and Kevin J. Murphy, "Compensation and Incentives: Practice vs. Theory," *Journal of Finance*, Vol. 43, No. 3, July 1988, pp. 593–616.

Beker, Victor A., and Esteban Albisu, "Raw Labor: Homogeneous or Heterogeneous?" April 29, 2010. As of February 3, 2017: http://papers.ssrn.com/sol3/papers.cfm?abstract_id=1597701

Biggs, Andrew, and Jason Richwine, *Comparing Federal and Private Sector Compensation*, Washington, D.C.: American Enterprise Institute, Economic Policy Working Paper 2011-02, rev. June 2011. As of February 3, 2017:
https://www.aei.org/wp-content/uploads/2011/10/AEI-Working-Paper-on-Federal-Pay-May-2011.pdf

Borjas, George J., *The Wage Structure and the Sorting of Workers into the Public Sector*, Cambridge, Mass.: National Bureau of Economic Research, Working Paper 9313, October 2002. As of February 3, 2017:
http://www.nber.org/papers/w9313

Bradley, David H., *Comparing Compensation for Federal and Private Sector Workers: An Overview*, Washington, D.C.: Congressional Research Service, July 30, 2012.

Brynjolfsson, Erik, and Adam Sanders, *Wired for Innovation: How Information Technology Is Reshaping the Economy*, Cambridge, Mass.: MIT Press, 2013.

Bureau of Economic Analysis, "Gross-Domestic-Product-(GDP)-by-Industry Data," web page, last updated October 16, 2016a. As of February 3, 2017:
http://www.bea.gov/industry/gdpbyind_data.htm

———, "Industry Economics Account Information Guide," web page, last updated October 16, 2016b. As of February 3, 2017:
http://www.bea.gov/industry/iedguide.htm

CGI Group, *Why Managed Services and Why Not Staff Augmentation? Ensuring Companies Derive the Most Value, Including Flexibility and Skill Access, from IT Service Providers*, Montreal, Quebec, 2015. As of February 3, 2017:
https://www.cgi.com/sites/default/files/white-papers/cgi-why-managed-services-why-not-staff-augmentation.pdf

CNN, "The Reagan Years: Reaganomics," 2001. As of February 3, 2017:
http://www.cnn.com/SPECIALS/2001/reagan.years/whitehouse/reaganomics.html

Coase, Ronald H., "The Nature of the Firm," *Economica*, Vol. 4, No. 16, November 1937, pp. 386–405.

Commission on Wartime Contracting in Iraq and Afghanistan, *Transforming Wartime Contracting: Controlling Costs, Reducing Risks*, final report to Congress, Washington, D.C., August 2011.

Contractor Manpower Reporting, *Contracting Officer Representative and Contracting Officer Technical Representative User Guide*, Contractor Manpower Reporting Application, version 3.6, undated(a).

———, *Contractor User Guide*, Contractor Manpower Reporting Application, version 3.6, undated(b).

———, *Full User Guide*, Contractor Manpower Reporting Application, version 3.6, undated(c).

———, *Resource Manager and Requiring Activity Manager User Guide*, Contractor Manpower Reporting Application, version 3.6, undated(d).

———, *Subcontractor User Guide*, Contractor Manpower Reporting Application, version 3.6, undated(e).

Cooper, Robin, and Robert S. Kaplan, "Make Cost Right: Make the Right Decisions," *Harvard Business Review*, September–October 1988, pp. 96–103.

Dobler, Donald W., and David N. Burt, *Purchasing and Supply Management: Text and Cases*, 6th ed., New York: McGraw-Hill, 1996.

Dunigan, Molly, *Victory for Hire: Private Security Companies' Impact on Military Effectiveness*, Stanford, Calif.: Stanford University Press, 2011.

Edwards, Chris, "Reducing the Costs of Federal Worker Pay and Benefits," *Downsizing the Federal Government*, September 20, 2016. As of February 3, 2017: http://www.downsizinggovernment.org/federal-worker-pay

Executive Office of the President, Bureau of the Budget, "Commercial-Industrial Activities of the Government Providing Products or Services for Governmental Use," Washington, D.C., Bulletin No. 55-4, January 15, 1955. As of February 3, 2017:
http://www.governmentcompetition.org/uploads/Bureau_of_the_Budget_Bulletin_55-4_January_15_1955.pdf

Falk, Justin, *Comparing the Compensation of Federal and Private Sector Employees*, Washington, D.C.: Congressional Budget Office, January 2012. As of February 3, 2017:
https://www.cbo.gov/sites/default/files/112th-congress-2011-2012/reports/01-30-FedPay_0.pdf

Gates, Susan M., and Albert A. Robbert, *Personnel Savings in Competitively Sourced DoD Activities: Are They Real? Will They Last?* Santa Monica, Calif.: RAND Corporation, MR-1117-OSD, 2000. As of February 3, 2017: http://www.rand.org/pubs/monograph_reports/MR1117.html

Gibbs, Michael, *Pay Competitiveness and Quality of Department of Defense Scientists and Engineers*, Santa Monica, Calif.: RAND Corporation, MR-1312-OSD, 2001. As of February 3, 2017:
http://www.rand.org/pubs/monograph_reports/MR1312.html

Glanz, James, and Alissa J. Rubin, "From Errand to Fatal Shot to Hail of Fire to 17 Deaths," *New York Times*, October 3, 2007a.

———, "Blackwater Shootings 'Murder,' Iraq Says," *New York Times*, October 8, 2007b.

Grasso, Valerie Bailey, *Defense Outsourcing: The OMB Circular A-76 Policy*, Washington, D.C.: Congressional Research Service, June 30, 2005.

Greider, William, "The Education of David Stockman," *Atlantic Monthly*, December 1981. As of February 3, 2017:
http://www.theatlantic.com/magazine/archive/1981/12/the-education-of-david-stockman/305760

Hamel, Gary, and C. K. Prahalad, *Competing for the Future*, Boston, Mass.: Harvard Business School Press, 1994.

Hoecht, A., and P. Trott, "Innovation Risks of Strategic Outsourcing," *Technovation*, Vol. 26, Nos. 5–6, May–June 2006, pp. 672–681.

James, Kay Coles, *A Fresh Start for Federal Pay: The Case for Modernization*, Washington, D.C.: Office of Personnel Management, April 2002. As of February 3, 2017:
http://archive.opm.gov/strategiccomp/whtpaper.pdf

Kelman, Steven, *Procurement and Public Management: The Fear of Discretion and the Quality of Government Performance*, Washington, D.C.: AEI Press, 1990.

Klitgaard, Robert, and Paul C. Light, eds., *High-Performance Government: Structure, Leadership, Incentives*, Santa Monica, Calif.: RAND Corporation, MG-256-PRGS, 2005. As of February 3, 2017:
http://www.rand.org/pubs/monographs/MG256.html

Leenders, Michiel R., Harold E. Fearon, Anna E. Flynn, and P. Fraser Johnson, *Purchasing and Supply Management*, 12th ed., New York: McGraw-Hill Irwin, 2002.

McIvor, Ronan, "What Is the Right Outsourcing Strategy for Your Process?" *European Management Journal*, Vol. 26, No. 1, February 2008, pp. 24–34.

Monczka, Robert M., Robert J. Trent, and Robert B. Handfield, *Purchasing and Supply Chain Management*, 2nd ed., Cincinnati, Ohio: South-Western College Pub., 2002.

Moore, Nancy Y., Laura H. Baldwin, Frank A. Camm, and Cynthia R. Cook, *Implementing Best Purchasing and Supply Management Practices: Lessons from Innovative Commercial Firms*, Santa Monica, Calif.: RAND Corporation, DB-334-AF, 2002. As of February 3, 2017:
http://www.rand.org/pubs/documented_briefings/DB334.html

Moore, Nancy Y., Clifford A. Grammich, and Robert Bickel, *Developing Tailored Supply Strategies*, Santa Monica, Calif.: RAND Corporation, MG-572-AF, 2007. As of February 3, 2017:
http://www.rand.org/pubs/monographs/MG572.html

Moore, Nancy Y., Clifford A. Grammich, Julie DaVanzo, Bruce J. Held, John Coombs, and Judith D. Mele, *Enhancing Small-Business Opportunities in the DoD*, Santa Monica, Calif.: RAND Corporation, TR-601-1-OSD, 2008. As of February 3, 2017:
http://www.rand.org/pubs/technical_reports/TR601-1.html

Moore, Nancy Y., Clifford A. Grammich, and Judith D. Mele, *Findings from Existing Data on the Department of Defense Industrial Base*, Santa Monica, Calif.: RAND Corporation, RR-614-OSD, 2014. As of February 3, 2017: http://www.rand.org/pubs/research_reports/RR614.html

Office of the Deputy Secretary of Defense, "Insourcing Contracted Services: Implementation Guidance," memorandum, May 28, 2009.

Office of Management and Budget, *Guidelines and Discount Rates for Benefit-Cost Analysis of Federal Programs*, Circular No. A-94, revised October 29, 1992.

———, *Performance of Commercial Activities*, Circular No. A-76, revised May 29, 2003.

Office of the Secretary of Defense, "Enterprise-Wide Contractor Manpower Reporting Application," memorandum, November 28, 2012.

Office of the Under Secretary of Defense, Comptroller, *National Defense Budget Estimates for FY 2017*, Washington, D.C., March 2016. As of February 3, 2017: http://comptroller.defense.gov/Portals/45/Documents/defbudget/fy2017/FY17_Green_Book.pdf

Office of the Under Secretary of Defense for Acquisition and Technology, *Report of the Defense Science Board Task Force on Outsourcing and Privatization*, Washington, D.C., August 1996. As of February 3, 2017: http://oai.dtic.mil/oai/oai?verb=getRecord&metadataPrefix=html&identifier=ADA316936

Public Law 107-107, National Defense Authorization Act for Fiscal Year 2002, December 28, 2001.

Public Law 110-161, Consolidated Appropriations Act, 2008, December 26, 2007.

Public Law 110-181, National Defense Authorization Act for Fiscal Year 2008, January 28, 2008.

Public Law 111-84, National Defense Authorization Act for Fiscal Year 2010, October 28, 2009.

Public Law 111-117, Consolidated Appropriations Act, 2010, December 16, 2009.

Public Law 111-259, Intelligence Authorization Act for Fiscal Year 2010, October 7, 2010.

Public Law 111-383, National Defense Authorization Act for Fiscal Year 2011, January 7, 2011.

Public Law 112-10, Department of Defense and Full-Year Continuing Appropriations Act, 2011, April 5, 2011.

Public Law 112-81, National Defense Authorization Act for Fiscal Year 2012, December 31, 2011.

Public Law 112-239, National Defense Authorization Act for Fiscal Year 2013, January 2, 2013.

Robbert, Albert A., Susan M. Gates, and Marc N. Elliot, *Outsourcing of DoD Commercial Activities: Impacts on Civil Service Employees*, Santa Monica, Calif.: RAND Corporation, MR-866-OSD, 1997. As of February 3, 2017: http://www.rand.org/pubs/monograph_reports/MR866.html

Rogin, Josh, "The Hidden Price of a Buying Spree," *CQ Weekly*, July 23, 2007.

Sherk, James, *Inflated Federal Pay: How Americans Are Overtaxed to Overpay the Civil Service*, Washington, D.C.: Heritage Foundation, CDA10-05, July 7, 2010. As of February 3, 2017:
http://www.heritage.org/research/reports/2010/07/inflated-federal-pay-how-americans-are-overtaxed-to-overpay-the-civil-service

Sizemore, Bill, and Joanne Kimberlin, "Blackwater, Part 4: When Things Go Wrong," *Virginian-Pilot*, July 26, 2006. As of February 3, 2017:
http://pilotonline.com/news/military/local/blackwater-part-when-things-go-wrong/article_1118107d-30d8-527f-869b-89eece61bfb5.html

Teece, David J., Gary Pisano, and Amy Shuen, "Dynamic Capabilities and Strategic Management," *Strategic Management Journal*, Vol. 18, No. 7, August 1997, pp. 509–533.

U.S. Department of Defense, *Time-and-Materials and Labor-Hour Contracts: The New Policies*, Washington, D.C., 2006a. As of February 3, 2017:
http://www.acq.osd.mil/dpap/dars/pgi/attachments/2006d030_overview.pdf

———, *Quadrennial Defense Review Report*, Washington, D.C., February 6, 2006b. As of February 3, 2017:
http://archive.defense.gov/pubs/pdfs/QDR20060203.pdf

———, *Resource Management Decision 802*, Washington, D.C., April 8, 2009.

———, *Product Support Business Case Analysis Guidebook*, Washington, D.C., April 2011. As of February 3, 2017:
https://acc.dau.mil/CommunityBrowser.aspx?id=452296

———, *FY14 CMRA PSC Rates and Factors*, Washington, D.C., 2014.

U.S. Department of Defense Directive 1100.4, *Guidance for Manpower Management*, February 12, 2005.

U.S. Department of Defense Instruction 1100.22, *Policy and Procedures for Determining Workforce Mix*, April 12, 2010.

U.S. Department of Defense Instruction 5000.74, *Defense Acquisition of Services*, January 5, 2016.

U.S. Department of Defense Instruction 7041.03, *Economic Analysis for Decision-Making*, September 9, 2015.

U.S. Department of Defense, Office of the Inspector General, *Contracts for Professional, Administrative, and Management Support Services*, Washington, D.C., Report No. D-2004-015, October 30, 2003.

———, *Independent Auditor's Report on Agreed-Upon Procedures for DoD Compliance with Service Contract Inventory Compilation and Certification Requirements for FY 2012*, Washington, D.C., Report No. DODIG-2014-114, September 17, 2014.

———, *Independent Auditor's Report on Agreed-Upon Procedures for DoD Compliance with Service Contract Inventory Compilation and Certification Requirements for FY 2013*, Washington, D.C., Report No. DODIG-2015-106, April 15, 2015.

U.S. Department of Defense, Office of Small Business Programs, "Small Business Program Goals," web page, undated. As of February 3, 2017:
http://www.acq.osd.mil/osbp/statistics/sbProgramGoals.shtml

U.S. Department of Labor, Executive Office of the President, and Office of Personnel Management, *Report on Locality-Based Comparability Payments for the General Schedule: Annual Report of the President's Pay Agent*, Washington, D.C., 2014. As of February 3, 2017:
https://www.opm.gov/policy-data-oversight/pay-leave/pay-systems/general-schedule/pay-agent-reports/2014report.pdf

U.S. General Accounting Office, *Outsourcing DOD Logistics: Savings Achievable but Defense Science Board's Projections Are Overstated*, Washington, D.C., NSIAD-98-48, December 1997. As of February 3, 2017:
http://www.gao.gov/products/NSIAD-98-48

———, *Defense Management: DOD Faces Challenges Implementing Its Core Competency Approach and A-76 Competitions*, Washington, D.C., GAO-03-818, July 2003. As of February 3, 2017:
http://www.gao.gov/assets/240/238966.pdf

U.S. General Services Administration, "System for Award Management," homepage, undated. As of February 3, 2017:
https://www.sam.gov/portal/SAM

———, *SAM Functional Data Dictionary*, version 7.0, Washington, D.C., July 17, 2014. As of February 3, 2017:
https://gsa.github.io/sam_api/sam/SAM_Functional_Data_Dictionary_v7_Public.pdf

———, *Federal Procurement Data System Product and Services Code Manual*, Washington, D.C., August 2015. As of February 3, 2017:
https://www.acquisition.gov/sites/default/files/page_file_uploads/PSC%20Manual%20-%20Final%20-%209%20August%202015_0.pdf

U.S. Government Accountability Office, *Contract Management: DOD Vulnerabilities to Contracting Fraud, Waste, and Abuse*, Washington, D.C., GAO-06-838R, July 2006. As of February 3, 2017:
http://www.gao.gov/products/GAO-06-838R

———, *Defense Acquisitions: Observations on the Department of Defense Service Contract Inventories for Fiscal Year 2008*, Washington, D.C., GAO-10-350R, January 2010a. As of February 3, 2017:
http://www.gao.gov/products/GAO-10-350R

———, *Contingency Contracting: Improvements Needed in Management of Contractors Supporting Contract and Grant Administration in Iraq and Afghanistan*, Washington, D.C., GAO-10-357, April 2010b. As of February 3, 2017:
http://www.gao.gov/products/GAO-10-357

———, *Defense Acquisitions: Further Action Needed to Better Implement Requirements for Conducting Inventory of Service Contract Activities*, Washington, D.C., GAO-11-192, January 2011. As of February 3, 2017:
http://www.gao.gov/products/GAO-11-192

———, *Defense Acquisitions: Further Actions Needed to Improve Accountability for DOD's Inventory of Contracted Services*, Washington, D.C., GAO-12-257, April 2012. As of February 3, 2017:
http://www.gao.gov/products/GAO-12-357

———, *Defense Acquisitions: Continued Management Attention Needed to Enhance Use and Review of DOD's Inventory of Contracted Services*, Washington, D.C., GAO-13-941, May 2013. As of February 3, 2017:
http://www.gao.gov/products/GAO-13-491

———, *Civilian Intelligence Community: Additional Actions Needed to Improve Reporting on and Planning for the Use of Contract Personnel*, Washington, D.C., GAO-14-204, January 2014a. As of February 3, 2017:
http://www.gao.gov/products/GAO-14-204

———, *Defense Acquisitions: Update on DOD's Efforts to Implement a Common Contractor Manpower Data System*, Washington, D.C., GAO-14-491R, May 2014b. As of February 3, 2017:
http://www.gao.gov/products/GAO-14-491R

———, *Defense Contractors: Additional Actions Needed to Facilitate the Use of DOD's Inventory of Contracted Services*, Washington, D.C., GAO-15-88, November 2014c. As of February 3, 2017:
http://www.gao.gov/products/GAO-15-88

———, *High-Risk Series: An Update*, Washington, D.C., GAO 15-290, February 2015a. As of February 3, 2017:
http://www.gao.gov/products/GAO-15-290

———, *DOD Inventory of Contracted Services: Actions Needed to Help Ensure Inventory Data Are Complete and Accurate*, Washington, D.C., GAO 16-46, November 2015b. As of February 3, 2017:
http://www.gao.gov/products/GAO-16-46

———, *Civilian and Contractor Workforces: Complete Information Needed to Assess DOD's Progress for Reductions and Associated Savings*, Washington, D.C., GAO 16-172, December 2015c. As of February 3, 2017:
http://www.gao.gov/products/GAO-16-172

U.S. House of Representatives, *Departments of Transportation and Housing and Urban Development, and Related Agencies Appropriations Act, 2010*, conference report to accompany H.R. 3288, December 8, 2009. As of February 3, 2017:
https://www.gpo.gov/fdsys/pkg/CRPT-111hrpt366/pdf/CRPT-111hrpt366.pdf

———, *National Defense Authorization Act for Fiscal Year 2016*, conference report to accompany H.R. 1735, September 2015.

U.S. House of Representatives, Committee on Appropriations, *Department of Defense Appropriations Bill, 2010*, Report 111-230 to accompany H.R. 3326, July 24, 2009. As of February 3, 2017:
https://www.gpo.gov/fdsys/pkg/CRPT-111hrpt230/pdf/CRPT-111hrpt230.pdf

U.S. House of Representatives, Committee on Armed Services, *National Defense Authorization Act for Fiscal Year 2006*, Report 109-89 to accompany H.R. 1815, May 20, 2005. As of February 3, 2017:
https://www.congress.gov/109/crpt/hrpt89/CRPT-109hrpt89.pdf

———, *National Defense Authorization Act for Fiscal Year 2011*, Report 111-491 to accompany H.R. 5136, May 21, 2010. As of February 3, 2017:
https://www.congress.gov/111/crpt/hrpt491/CRPT-111hrpt491.pdf

U.S. Senate, Committee on Appropriations, "Examining the Effectiveness of U.S. Efforts to Combat Waste, Fraud, Abuse, and Corruption in Iraq," Senate Hearing 110-673, Washington, D.C., March 11 and July 23, 2008. As of February 3, 2017:
https://www.gpo.gov/fdsys/pkg/CHRG-110shrg43280/html/CHRG-110shrg43280.htm

U.S. Senate, Committee on Armed Services, *National Defense Authorization Act for Fiscal Year 2002*, Report 107-62 to accompany S. 1416, September 12, 2001. As of February 3, 2017:
https://www.gpo.gov/fdsys/pkg/CRPT-107srpt62/pdf/CRPT-107srpt62.pdf

U.S. Senate, Committee on Governmental Affairs, *Federal Activities Inventory Reform Act of 1998*, Report 105-269 to accompany S. 314, July 28, 1998. As of February 3, 2017:
https://www.gpo.gov/fdsys/pkg/CRPT-105srpt269/html/CRPT-105srpt269.htm

U.S. Small Business Administration, *Government Contracting 101: A Guide for Small Businesses, Part 1—Small Business Contracting Programs, Supplemental Workbook*, January 2012. As of February 3, 2017:
https://www.sba.gov/sites/default/files/files/Work%20book%20gc%20101%20part%201.pdf

University of Denver, Private Security Monitor, "Articles, Reports, and Statistics," web page, undated. As of February 3, 2017:
http://psm.du.edu/articles_reports_statistics/data_and_statistics.html

Walker, David M., Comptroller General of the United States, *Results-Oriented Government: Shaping the Government to Meet 21st Century Challenges*, statement before the Subcommittee on Civil Service and Agency Organization, Committee on Government Reform, U.S. House of Representatives, Washington, D.C.: U.S. General Accounting Office, GAO-03-1168T, September 17, 2003. As of February 3, 2017:
http://www.gao.gov/products/GAO-03-1168T

Wernerfelt, Birger, "A Resource Based View of the Firm," *Strategic Management Journal*, Vol. 5, No. 2, April–June 1984, pp. 171–180.